WHICH SIDE OF HISTORY?

WHICH SIDE OF HISTORY?

HOW TECHNOLOGY IS RESHAPING DEMOCRACY AND OUR LIVES

EDITED BY JAMES P. STEYER

CHRONICLE PRISM

Library of Congress Cataloging-in-Publication Data available.
ISBN 978-1-7972-0516-8
Manufactured in the United States of America.

Interior design by Sara Schneider.
Typesetting by Happenstance Type-O-Rama.
Cover design by Kelley Galbreath.

10 9 8 7 6 5 4 3 2

CHRONICLE PRISM

Chronicle Prism is an imprint of Chronicle Books LLC, 680 Second Street,
San Francisco, California 94107
www.chronicleprism.com

FOR MY LOVELY WIFE, LIZ. A GREAT PERSON AND A GREAT MOM, WHO IS DEFINITELY ON THE RIGHT SIDE OF HISTORY.

JAMES P. STEYER

CONTENTS

PART 3

A THREAT TO DEMOCRACY?

PART 4

WHERE BIG TECH WENT WRONG

PART 5

TECHNOLOGY AND RACE

PART 6

DOING GOOD, NOT EVIL

INTRODUCTION

JAMES P. STEYER

James P. Steyer is founder and chief executive officer of Common Sense Media—the nation's leading nonpartisan organization dedicated to improving media and technology choices for kids and families—and longtime award-winning professor at Stanford University.

We are living in a truly remarkable time. Our lives and entire society have been transformed by a technology revolution and its 24/7 influence on so many aspects of our reality. The COVID-19 pandemic and the nationwide movement for racial justice have made it even clearer how essential these tools are for our daily lives. When I wrote my last book, *Talking Back to Facebook*, in 2012, my goals were simple enough: first, to highlight the extraordinary impact that the internet and technology companies were having on the lives of kids, families, and the fabric of our social and emotional relationships. And second, to give parents a few tips on how to raise kids in this new era of smartphones and social media. Little did I know that a mere eight years later, most thoughtful people would indeed be "talking back to Facebook"—and that the grip of smartphones and social media would impact and at times threaten not only our children and family relationships but virtually everything we hold dear.

Silicon Valley has clearly followed through on its promise to change the world on a vast scale—not necessarily for the better. Bot by bot and tweet by tweet, the ghastly circus of the 2016 US election hacking, as well as the spiraling descent of our public discourse, have put the very pillars of our democracy at risk. Smartphones and social media have chummed a surveillance and attention economy—a virtual arms race for your and my attention—designed to invade our privacy, monetize our secrets, and steal every waking moment of our lives. Not by accident, we are lonelier, sadder, more anxious, and more divided as a result. Perhaps most disturbing, the most vulnerable laboratory animals for tech's sweeping, unregulated, and truly pioneering social experiment have been an entire generation of innocent, unwitting kids, including my own four children. More than any other segment of society, our kids have been in the crosshairs of the technology revolution. They are the ones who will most profoundly experience its benefits, disruptions, and consequences for our future.

Facebook, according to its founders, was supposed to chronicle our lives and bring us together in unprecedented ways of "frictionless sharing." Now it more often serves to chronicle and accelerate our collective slide toward the abyss. No company has ever known so much or learned so little in such a short time. Facebook began the last decade by entering into a major consent decree with the Federal Trade Commission for widespread privacy violations. Sadly, it ended the decade with the FTC imposing a record $5 billion fine—widely criticized as too small—for breaching that same agreement. Facebook can name all your friends, track your location, and predict your next purchase, but it also served as the main platform for Russian interference in our 2016 election and fatuously claimed not to realize that was happening.

Inevitably, the company whose mantra explicitly set out to "move fast and break things" finally broke its own industry's darkest secrets wide open. The Cambridge Analytica scandal has forced a reckoning with the devil's bargains that our society has chosen to ignore for much too long.

A colleague and I met with top executives at Facebook in March 2018, as breaking news of Cambridge Analytica caused the company's market capitalization to plunge $36 billion in one day. We expected our friends at Facebook to be contrite. Instead, its leaders rejected our suggestion to abandon political advertising. They even attacked us for criticizing the company's efforts to market a version of Facebook to kids as young as six years old. No company so clearly determined to choose the wrong side of history can be trusted with our future.

Facebook is by no means alone in the damage it has wrought to our society, but it is hardly the only key actor in this global saga. A number of other companies have disrupted, and in many cases fundamentally altered, critical aspects of our lives and our broader society—some for better and some for worse.

That said, the counterrevolution has already begun. Each data breach, privacy violation, and disturbing new means of exploiting personal data erodes confidence in the tech industry's intentions. A 2019 Pew Research Center survey found that only 50 percent of Americans now believe technology companies have a positive impact, down from 71 percent in 2015—a truly remarkable shift in a short time.

We all now find ourselves at a critical inflection point. Will we continue to trust Big Tech to make the world a better place, when—if we can look up from our phones long enough—our own eyes tell us the world seems headed full speed the other way? Or will we recognize that when an industry has unlimited capacity for good or evil, we must hold it accountable and not leave the most powerful companies on the face of the earth to their own devices? Each of us must answer this fundamental question, whether we are a citizen, parent, student, elected official, or tech CEO.

Which Side of History? frames these critical issues and presents the reflections, warnings, and recommendations of some of the most influential thought leaders of our time. These essays explore technology's key benefits as well as harms, and they offer solutions and suggestions for how to address its damages and dangers before it is too late. From inside and outside the tech industry, these powerful voices—of journalists, engineers, entrepreneurs, novelists, filmmakers, business leaders, scholars, and researchers—provide a 360-degree view of the present and future impact of technology on our democracy, our society, and our lives. Their insights can help inform the debates, decisions, and policies that will guide the direction of our country for years to come.

The book is arranged in six parts that focus on some central themes:

Part 1, "History Is Watching," examines the considerable damage that unregulated technology is already doing, from the rise of tech addiction and social isolation to the death of privacy and the breakdown of democratic norms. Tech troubles, fortunately, have finally sparked a genuine crisis of conscience within the industry. Top college graduates from schools like Stanford, where I've been a professor for thirty years, once flocked to tech companies; now, many think twice. Former reddit CEO Ellen Pao ("Tech, Heal Thyself," p. 31) warns that tech companies need more diversity and stronger values, and leaders must "measure and hold ourselves accountable for harm to employees, partners, consumers, the environment, neighbors, and the public community." And a few truly brave tech leaders are challenging their peers to save the industry from itself.

My good friend, the irrepressible, visionary Salesforce founder and CEO Marc Benioff ("We Need a New Capitalism, Based on Trust," p. 41) calls Facebook "the new cigarettes" and warns that "tech companies can no longer wash their hands of what people do with our products." The gifted actor and activist Sacha Baron Cohen ("The Greatest Propaganda Machine in History," p. 3) argues persuasively for regulation and responsibility: "It's time to finally call these companies what they really are—the largest publishers in history. They should abide by basic standards and practices just like newspapers, magazines, and TV news do every day."

Part 2, "How Tech Is Hurting Kids," explores the myriad ill effects of technology use among young people. According to a 2019 state-of-the-art research study by Common Sense Media, American teens log an average of seven and a half hours per day of screen time, not including time spent using screens for school or homework. The average eight- to twelve-year-old American child watches nearly five hours per day of screen-based entertainment and media, again not including time spent using screens for school.

The impact of this exposure is enormous. The 24/7 technology and media experience has enormous implications for the social, emotional, and cognitive development of our kids and has also contributed to a unique and growing mental health crisis among youth. Noted psychologist Madeline Levine ("Kids Interrupted: How Social Media Derails Adolescent Development," p. 77) documents damning evidence of the toxic effects of social media, including rising rates of isolation, anxiety, and depression. "Given the runaway dependence of most teenagers on social media sites," she writes, "we are at a point of much-needed reevaluation around the role and impact of social media on our kids' lives as well as our own." Leading policy expert Bruce Reed and I ("Why Section 230 Hurts Kids—and What to Do about It," p. 94) argue for taking away the social media platforms' shield of blanket immunity for the massive amounts of content that they host and distribute.

In part 3, "A Threat to Democracy?," writers consider the impact of Russia's use of social media to manipulate global politics. While America slept, Russian troll factories effortlessly manipulated the 2016 US election, the Brexit election in the UK, and others. Thankfully, prominent figures in both political parties—though, sadly, not all—are trying to harden democracy's defenses. For Senator Mark Warner ("The Assault on Civil Discourse and an Informed Electorate," p. 136), the consequences are clear: "The misuse of technology threatens our democratic systems, the robustness of our economy, and our national security." Similarly, for voting rights activist LaTosha Brown ("Using Technology to Defeat Democracy," p. 133), technology has been a tool to suppress the basic democratic rights of voters of color. Meanwhile, the White House and Congress have largely watched the glories and transgressions of Big Tech from the sidelines. Their MIA behavior is simply unconscionable and has had extraordinary consequences for our nation. The more powerful and omniscient the tech industry becomes, the more desperately we need an informed, functioning national government to look out for us. We have seen how much technology can do to put our very democratic institutions on the ropes. It is well past time for our elected leaders to get up off the mat, do their jobs, and hold tech companies accountable to the public good.

In part 4, "Where Big Tech Went Wrong," Thrive Global CEO Arianna Huffington ("Technology Can Augment Our Humanity or Consume It," p. 178) points out that "conversations about technology's effects have been happening for years, but the revelations about how its use helped undermine the election pushed those conversations into the collective consciousness." The intense spotlight on tech has also laid bare the business model behind the industry's sins. In exchange for free apps and other alluring offers, consumers trade away their data and personal privacy, enabling companies to track their every move and microtarget their every desire. People are beginning to realize what a bad trade that turned out to be. According to Ken Auletta, the renowned media and technology author ("Mad Men and Math Men," p. 195), the personal data these companies use to predict our behavior is the "holy grail" for advertisers and generates roughly 80 to 90 percent of revenue at tech giants like Google and Facebook. In the words of the widely respected *New York Times* columnist and writer Thomas Friedman ("We're All Connected but No One's in Charge," p. 173), we now "feel beat up by the same platforms and technologies that had enriched, empowered, and connected our lives."

You simply can't talk about technology today without addressing the systemic problems of racial injustice and inequality in our society. **In part 5, "Technology and Race,"** we are fortunate to have some of the most thoughtful, insightful voices in the world addressing the intersection of technology and many of these vital issues. Powerful essays–by my friends Geoffrey Canada ("Closing the Digital Divide," p. 215), Michelle Alexander ("The Newest Jim Crow," p. 218), and Theodore M. Shaw ("Technology, Inclusiveness, Structural Racism, and Silicon Valley," p. 223), as well as the widely respected scholar and author Ruha Benjamin ("The New Jim Code," p. 211)–make clear that the major tech companies will inevitably be judged by their impact, for better or worse, on racial justice and inequality in America. History will certainly be watching them.

Finally, part 6, "Doing Good, Not Evil," raises a range of tough questions: What are we going to do about this? Whom are we going to hold accountable? What future do we really want to leave for our children and future generations? An industry bent on disruption is just getting warmed up, so might artificial intelligence and the "Fourth Industrial Revolution" even alter what it means to be human?

Across the United States and Europe, answers are slowly emerging. Attorneys general in nearly every US state have launched antitrust investigations into Facebook and other major tech companies. The European Union has enacted sweeping privacy protections and empowered regulators to hold tech companies accountable for misuse of personal data or their market power. According to early Facebook investor Roger McNamee ("Making Internet Platforms Accountable," p. 267), "Platforms are more powerful than governments" and must be reined in. And as the leading private-equity CEO James G. Coulter ("The Change in the Nature of Change," p. 292) suggests, "It is time to build a coherent governmental structure to help our democracy address the challenges of data protection, data ownership, privacy, digital citizenship, and platform regulation." In the end, humankind will not surrender our rights, our privacy, our democracy, and our very humanity without a fight.

The 2020 COVID-19 pandemic illuminated the many ways technology can bring us together when we're forced to be apart. When millions of students and parents saw their homes suddenly turned into schools, Common Sense Media sprang into action to create and curate Wide Open School, the largest collection of free online learning experiences for kids. One of the twenty-five organizations to help launch that effort was Khan Academy. Educator Sal Khan ("How Technology Can Humanize Education," p. 249) points out how technology can be used for good, allowing people "to move into new professions with mastery and confidence" through personalized, online education. Companies like Google,

Apple, and Zoom also stepped forward in a big way, recognizing the importance of their platforms for the education of millions of American students.

If we're looking for hope, there may be no brighter sign than in the tech industry's backyard of California. To the surprise of many, my home state has become the front lines of the battle to hold Big Tech to account and to leverage the power of its influence in extraordinarily positive ways. A number of tech leaders are partnering with those of us who have been standing up for kids, privacy, and digital well-being all along.

One of those organizations is Common Sense Media, a nonprofit I founded in 2003 to give parents and educators the tools to help kids navigate the digital world. Today, well more than a hundred million parents and teachers regularly turn to Common Sense for reviews, advice, and advocacy. On their behalf, we have made significant strides at the state and federal level to hold technology companies to a far higher standard. We helped rewrite the national Children's Online Privacy Protection Act (COPPA), and we introduced cutting-edge California legislation like the 2014 "eraser button" law, which lets young people wipe away whatever mistaken images and messages they may have posted online as teens. For years, we have been working to close the "Homework Gap" for young people who lack access to the internet.

In 2018, Common Sense conceived of and cosponsored the landmark California Consumer Privacy Act (CCPA), which became the de facto law of the land in the United States at the beginning of 2020. The CCPA gives all consumers the right to find out what personal data is collected about them and with whom that data is shared; the right to say no to the sale of their data; and the right to ask a business to delete that information. This law also requires businesses to obtain affirmative consent from consumers under the age of sixteen, who must opt in before their personal information can be sold.

When Common Sense Media pushed for this landmark legislation, all the experts and insiders said we didn't stand a chance. Everyone—including the tech industry itself—believed that Big Tech had the California legislature in their pocket. So many tech lobbyists roamed the halls in Sacramento that it felt like trying to regulate the railroads in the 1890s. But people are realizing that Big Tech has built a massive empire, sometimes at our personal expense—and most of us don't like what it means for our privacy, for our democracy and security, or for our children. When we beat the odds and finally brought this legislation to the floor of the California legislature, public support was so overwhelming for CCPA that not a single legislator, in either party, voted against it. So, yes, we have reason to be optimistic.

In the face of the tech industry's wealth, power, and influence, some may despair that we're doing too little, too late. In the wake of the COVID-19 pandemic, regulatory pressure stalled, and internet platforms looked to emerge bigger, stronger, and more powerful than ever. But the good news is that some of the most powerful tech companies stepped forward to protect the interests of kids, families, and schools. The lesson is clear. The real future of the technology industry depends on living up to its original promise, to foster connection—not to make the world a shallower, creepier, and more divided place. Put simply, it is time for all of us to make clear which side of history we're on.

PART 1

HISTORY IS WATCHING

The Greatest Propaganda Machine in History

SACHA BARON COHEN

Sacha Baron Cohen is a British actor, comedian, screenwriter, director, and film producer best known for fictional, satirical characters he has created and portrayed, including Ali G, Borat Sagdiyev, Brüno Gehard, and Admiral General Aladeen.

Some critics have said that my comedy, at times, risks reinforcing negative stereotypes. But the truth is, I've been passionate about challenging bigotry and intolerance throughout my life. As a teenager in the United Kingdom, I marched against the fascist National Front and to abolish apartheid. As an undergraduate, I traveled around America and wrote my thesis about the civil rights movement, with the help of the archives of the Anti-Defamation League (ADL). And as a comedian, I've tried to use my characters to get people to let down their guard and reveal what they actually believe, including their own prejudices.

I admit, there was nothing particularly enlightening about me—as Borat from Kazakhstan, the first fake news journalist—running naked through a conference of mortgage brokers. But when Borat was able to get an entire bar in Arizona to sing "Throw the Jew down the well," it did reveal people's indifference to anti-Semitism. When—as Brüno, the gay fashion reporter from

Austria—I started kissing a man in a cage fight in Arkansas, nearly starting a riot, it showed the violent potential of homophobia. And when—disguised as an ultrawoke developer—I proposed building a mosque in one rural community, prompting a resident to proudly admit, "I am racist, against Muslims," it showed the acceptance of Islamophobia.

Today around the world, demagogues appeal to our worst instincts. Democracy, which depends on shared truths, is in retreat, and autocracy, which depends on shared lies, is on the march. Hate crimes are surging, as are murderous attacks on religious and ethnic minorities. All this hate and violence is being facilitated by a handful of internet companies that amount to the greatest propaganda machine in history.

Think about it. Facebook, YouTube, Google, Twitter, and others reach billions of people. The algorithms these platforms depend on deliberately amplify stories that appeal to our baser instincts and trigger outrage and fear. It's why fake news outperforms real news—because studies show that lies spread faster than truth.

In their defense, these social media companies have taken some superficial steps to reduce hate and conspiracies on their platforms, but it's time to finally call these companies what they really are—the largest publishers in history. They should abide by basic standards and practices just like newspapers, magazines, and TV news do every day.

Publishers can be sued for libel, people can be sued for defamation. I've been sued many times! But social media companies are largely protected from liability for the content their users post—no matter how indecent it is—by Section 230 of, get ready for it, the Communications Decency Act. Absurd!

People should not be targeted, harassed, and murdered because of who they are, where they come from, who they love, or how they pray. If we prioritize truth over lies, tolerance over prejudice, empathy over indifference, and experts over ignoramuses—then maybe we can stop the greatest propaganda machine in history, save democracy, and still protect free speech and free expression.

Be Paranoid

KARA SWISHER

Kara Swisher is an editor-at-large at Recode and is a contributing opinion writer on technology for the New York Times.

Only the paranoid survive.

That was, of course, the motto made famous by Intel's legendary founder and former chief executive, Andy Grove, who later turned the line into a book that was actually about being hypervigilant as inevitable "crisis points" occur at your company.

It's still a good piece of advice, but these days it seems as though there is an entirely new way of reading that line when it comes to a different issue in tech: the surveillance economy that continues to spread like a virus worldwide, even as consumers are less aware than ever of its implications.

That includes me, who should know better. I have two-factor authentication. I cover my camera lens on my computer. I redo all my security settings regularly. I am wary of—you might even say mean about—various consumer abuses by giant social media companies, search behemoths, and testosterone-jacked e-commerce companies.

Still, as much as I know about tech, I'm often lazy and use its tools without care, even as each day seems to bring new headlines about privacy incursions sometimes done for commercial reasons, sometimes for malevolent ones, and sometimes just as a result of tech's latest changes. Privacy has been losing badly, as users have become the online equivalent of cheap dates to these giant tech companies. We trade the lucrative digital essence of ourselves for much less in the form of free maps or nifty games or compelling communications apps.

We're digitally sloppy, even if it can be very dangerous, as evidenced by a disturbing *New York Times* story about an Emirati secure messaging app called ToTok, which is used by millions across the Middle East and has also recently become one of the most downloaded in the United States.

The name was obviously used to place the app adjacent to the hugely popular TikTok, already under scrutiny by American officials because of its Chinese origins and possible link to the Beijing government. In the case of ToTok, according to the *Times* report, it turns out that it is a spy tool "used by the government of the United Arab Emirates to try to track every conversation, movement, relationship, appointment, sound, and image of those who install it on their phones."

The app's skein of developers is opaque, but apparently it is controlled by a sinister-sounding company linked to the Emirate government called DarkMatter. (Yes, that is actually its name, akin to calling the villain in a movie Mr. Really Bad Guy.)

After the *Times* inquiry, Google and Apple, US tech giants that are the prime distributors of apps worldwide, removed ToTok from their online stores. But the damage was done—and by the users themselves.

"You don't need to hack people to spy on them if you can get people to willingly download this app to their phone," said Patrick

Wardle, who did a forensic analysis for the *Times*, in the report. "By uploading contacts, video chats, location, what more intelligence do you need?"

Indeed, anyone who wants to spy needs very little, as all of us continue availing ourselves of tech's many wonders while promiscuously shedding our data.

That much was clear in the eye-opening investigation of smartphones by *Times Opinion* called "One Nation Tracked." The *Opinion* report was even more dire than the ToTok story: One data set of twelve million phones with fifty billion location pings from a basic location-data company showed clearly that there is no such thing as privacy. At all. Ever. Not on the beaches of Southern California, not at the Pentagon, not at the White House.

"Now, as the decade ends, tens of millions of Americans, including many children, find themselves carrying spies in their pockets during the day and leaving them beside their beds at night—even though the corporations that control their data are far less accountable than the government would be," noted the report, which included a look at how to track President Trump, the citizens of Pasadena, and protesters in Hong Kong, as well as how to try to stop it all. This is what freaked me out enough to go back and tighten the security on my own phone.

Yes, it's up to us to protect ourselves, since there are no federal laws that actually do it. Europe has been far ahead on privacy with its 2016 General Data Protection Regulation (which was implemented in 2018), and things will finally start to change for Californians like me as the 2018 California Consumer Privacy Act goes into effect.

The law will bring some relief, since it will give citizens of the state more control over personal data: We will know what is being collected and where it is sold, and we will even have the right to ask for such data to be deleted. It also adds special protections for minors, prohibiting the sale of personal information of those under sixteen years old.

California's law will become the de facto law of the land on privacy until the federal government acts, which is a long way from happening. Promises that bills will be rolled out in the House and Senate came and went in 2019.

The record so far is not encouraging. Which is why I'll opt out of waiting and keep fending off my app stalkers by myself.

Our world, at the moment, feels very Yeats-y—things falling apart, the center not holding, anarchy loosed, drowned innocence, a lack of conviction from the best, and of course, endless loud-mouthery from the worst.

But despair not because what's coming in the next few years might be a lot better than you expect, especially now that our outsize expectations for tech have been leveled and fanboy tendencies toward technology companies have been tamped down.

We haven't quite dealt with all of the repercussions of tech's domination of the past decade—there will be regulations, a lot of tech is still addictive, and digital hate will continue to travel halfway around the world before the truth gets out of bed—but there are some big, positive ideas that I think you will hear a lot more about in the coming years.

There are opportunities to create new forms of communication that give the advantage to users—by strictly enforcing behavior standards and eliminating anonymity, and most of all, with advertising-based business plans that are not predicated on taking advantage of our personal data.

There is yet another opportunity here to push for design ethics, a movement that I think will gain traction as we all assess what our dives into digital have done to humanity. While our tech devices have, on the whole, been good for most people, there is a true business opportunity in making them work more efficiently and without a reliance on addiction. Whether we move toward more intuitively created tech that surrounds us or that incorporates into our bodies (yes, that's coming), I am going to predict that carrying around a device in our hand and staring at it will be a thing of the past by 2030. Like the electrical grid we rely on daily, most tech will become invisible.

That's right, I am calling it now: There will be an internet in the future that stops screaming at us.

Only the paranoid survive, for sure, but so do the patient.

From the New York Times.

The Known Unknown

SHOSHANA ZUBOFF

Shoshana Zuboff is the author of The Age of Surveillance Capitalism *and is professor emerita at Harvard Business School.*

The debate on privacy and law at the Federal Trade Commission was unusually heated that day. Tech industry executives "argued that they were capable of regulating themselves and that government intervention would be costly and counterproductive." Civil libertarians warned that the companies' data capabilities posed "an unprecedented threat to individual freedom." One observed, "We have to decide what human beings are in the electronic age. Are we just going to be chattel for commerce?" A commissioner asked, "Where should we draw the line?" The year was 1997.

The line was never drawn, and the executives got their way. Twenty-three years later the evidence is in. The fruit of that victory was a new economic logic that I call "surveillance capitalism." Its success depends upon one-way-mirror operations engineered for our ignorance and wrapped in a fog of misdirection, euphemism, and mendacity. It rooted and flourished in the new spaces of the internet, once celebrated by surveillance capitalists as "the world's largest ungoverned space." But power fills a void, and those once-wild spaces are no longer ungoverned. Instead, they are owned and operated by private surveillance capital and governed by its iron laws.

The rise of surveillance capitalism over the last two decades went largely unchallenged. "Digital" was fast, we were told, and stragglers would be left behind. It's not surprising that so many of us rushed to follow the bustling White Rabbit down his tunnel into a promised digital Wonderland where, like Alice, we fell prey to delusion. In Wonderland, we celebrated the new digital services as free, but now we see that the surveillance capitalists behind those services regard us as the free commodity. We thought that we were searching Google, but now we understand that Google searches us. We assumed that we were using social media to connect, but we have learned that connection is how social media uses us. We barely questioned why our new TV or mattress had a privacy policy, but we've begun to understand that "privacy" policies are actually surveillance policies.

Like our forebears who called the automobile a "horseless carriage" because they could not reckon with its true dimension, we regarded the internet platforms as "bulletin boards" where anyone could pin a note. Congress cemented this delusion in a statute, Section 230 of the 1996 Communications Decency Act, giving those companies immunity from liability and absolving them of the obligations that adhere to "publishers" or even to "speakers."

Only repeated crises have taught us that these platforms are not bulletin boards but hypervelocity global bloodstreams into which anyone may introduce a dangerous virus without a vaccine. This is how Facebook's chief executive, Mark Zuckerberg, could legally refuse to remove a faked video of Speaker of the House Nancy Pelosi and later double down on this decision, announcing that political advertising would not be subject to fact-checking.

All of these delusions rest on the most treacherous hallucination of them all: the belief that privacy is private. We have imagined that we can choose our degree of privacy with an individual calculation in which a bit of personal information is traded for valued services—a reasonable quid pro quo. For example, when Delta Air Lines piloted a biometric data system at the Atlanta airport, the company reported that of nearly twenty-five thousand customers who traveled there each week, 98 percent opted into the process, noting that "the facial recognition option is saving an average of two seconds for each customer at boarding, or nine minutes when boarding a wide-body aircraft."

In fact, the rapid development of facial recognition systems reveals the public consequences of this supposedly private choice. Surveillance capitalists have demanded the right to take our faces wherever they appear—on a city street or a Facebook page. The *Financial Times* reported that a Microsoft facial recognition training database of ten million images plucked from the internet without anyone's knowledge and supposedly limited to academic research was employed by companies like IBM and state agencies that included the United States and Chinese militaries. Among these were two Chinese suppliers of equipment to officials in Xinjiang, where members of the Uighur community live in open-air prisons under perpetual surveillance by facial recognition systems.

Privacy is not private because the effectiveness of these and other private or public surveillance and control systems depends upon the pieces of ourselves that we give up—or that are secretly stolen from us.

Our digital century was to have been democracy's golden age. Instead, we enter its third decade marked by a stark new form of social inequality best understood as "epistemic inequality." It recalls a pre-Gutenberg era of extreme asymmetries of knowledge and the power that accrues to such knowledge, as the tech giants seize control of information and learning itself. The delusion of "privacy as private" was crafted to breed and feed this unanticipated social divide. Surveillance capitalists exploit the widening inequity of knowledge for the sake of profits. They manipulate the economy, our society, and even our lives with impunity, endangering not just individual privacy but democracy itself. Distracted by our delusions, we failed to notice this bloodless coup from above.

The belief that privacy is private has left us careening toward a future that we did not choose because it failed to reckon with the profound distinction between a society that insists upon sovereign individual rights and one that lives by the social relations of the one-way mirror. The lesson is that privacy is public—it is a collective good that is logically and morally inseparable from the values of human autonomy and self-determination upon which privacy depends and without which a democratic society is unimaginable.

Still, the winds appear to have finally shifted. A fragile new awareness is dawning as we claw our way back up the rabbit hole toward home. Surveillance capitalists are fast because they seek neither genuine consent nor consensus. They rely on psychic numbing and messages of inevitability to conjure the helplessness, resignation, and confusion that paralyze their prey. Democracy is slow, and that's a good thing. Its pace reflects the tens of millions of conversations that occur in families; among neighbors, coworkers, and friends; and within communities, cities, and states, which are gradually stirring the sleeping giant of democracy to action.

These conversations are occurring now, and there are many indications that lawmakers are ready to join and to lead. This third decade is likely to decide our fate. Will we make the digital future better, or will it make us worse? Will it be a place that we can call home?

Epistemic inequality is not based on what we can earn but rather on what we can learn. It is defined as unequal access to learning imposed by private commercial mechanisms of information capture, production, analysis, and sales. It is best exemplified in the fast-growing abyss between what we know and what is known about us.

Twentieth-century industrial society was organized around the "division of labor," and it followed that the struggle for economic equality would shape the politics of that time. Our digital century shifts society's coordinates from a division of labor to a "division of learning," and it follows that the struggle over access to knowledge and the power conferred by such knowledge will shape the politics of our time.

The new centrality of epistemic inequality signals a power shift from the ownership of the means of production, which defined the politics of the twentieth century, to the ownership of the production of meaning. The challenges of epistemic justice and epistemic rights in this new era are summarized in three essential questions about knowledge, authority, and power: *Who knows? Who decides who knows? Who decides who decides who knows?*

During the last two decades, the leading surveillance capitalists—Google, later followed by Facebook, Amazon, and Microsoft—helped to drive this societal transformation while simultaneously ensuring their ascendance to the pinnacle of the epistemic hierarchy. They operated in the shadows to amass huge knowledge monopolies by taking without asking, a maneuver that every child recognizes as theft. Surveillance capitalism begins by unilaterally staking a claim to private human experience as free raw material for translation into behavioral data. Our lives are rendered as data flows.

Early on, it was discovered that, unknown to users, even data freely given harbors rich predictive signals, a surplus that is more than what is required for service improvement. It isn't only what you post online, but whether you use exclamation points or the color saturation of your photos; not just where you walk but the stoop of your shoulders; not just the identity of your face but the emotional states conveyed by your "microexpressions"; not just what you like but the pattern of likes across engagements. Soon this behavioral surplus was secretly hunted and captured, claimed as proprietary data.

The data are conveyed through complex supply chains of devices, tracking and monitoring software, and ecosystems of apps and companies that specialize in niche data flows captured in secret. For example, testing by the *Wall Street Journal* showed that Facebook receives heart-rate data from the Instant Heart Rate: HR Monitor, menstrual cycle data from the Flo Period & Ovulation Tracker, and data that reveal interest in real estate properties from Realtor.com—all of it without the user's knowledge.

These data flows empty into surveillance capitalists' computational factories, called "artificial intelligence," where they are manufactured into behavioral predictions that are about us, but they are not for us. Instead, they are sold to business customers in a new kind of market that trades exclusively in human futures. Certainty in human affairs is the lifeblood of these markets, where surveillance capitalists compete on the quality of their predictions. This is a new form of trade that birthed some of the richest and most powerful companies in history.

In order to achieve their objectives, the leading surveillance capitalists sought to establish unrivaled dominance over the 99.9 percent of the world's information now rendered in digital formats that they helped to create. Surveillance capital has built most of the world's largest computer networks, data centers, servers, undersea transmission cables, advanced microchips, and frontier machine intelligence, igniting an arms race for the ten thousand or so specialists on the planet who know how to coax knowledge from these vast new data continents.

With Google in the lead, the top surveillance capitalists seek to control labor markets in critical expertise, including data science and animal research, elbowing out competitors such as startups, universities, high schools, municipalities, established corporations in other industries, and less wealthy countries. In 2016, 57 percent of American computer science PhD graduates took jobs in industry, while only 11 percent became tenure-track faculty members. It's not just an American problem. In Britain, university administrators contemplate a "missing generation" of data scientists. A Canadian scientist laments, "The power, the expertise, the data are all concentrated in the hands of a few companies."

Google created the first insanely lucrative markets to trade in human futures, what we now know as online targeted advertising, based on their predictions of which ads users would click. Between 2000, when the new economic logic was just emerging, and 2004, when the company went public, revenues increased by 3,590 percent. This startling number represents the "surveillance dividend." It quickly reset the bar for investors, eventually driving startups, app developers, and established companies to shift their business models toward surveillance capitalism. The promise of a fast track to outsized revenues from selling human futures drove this migration first to Facebook, then through the tech sector, and now throughout the rest of the economy to industries as disparate as insurance, retail, finance, education, health care, real estate, entertainment, and every product that begins with the word "smart" or service touted as "personalized."

Even Ford, the birthplace of the twentieth-century mass-production economy, is on the trail of the surveillance dividend, proposing to meet the challenge of slumping car sales by reimagining Ford vehicles as a "transportation operating system." As one analyst put it, Ford "could make a fortune monetizing data. They won't need engineers, factories, or dealers to do it. It's almost pure profit."

Surveillance capitalism's economic imperatives were refined in the competition to sell certainty. Early on it was clear that machine intelligence must feed on volumes of data, compelling economies of scale in data extraction. Eventually it was understood that volume is necessary but not sufficient. The best algorithms also require varieties of data—economies of scope. This realization helped drive the "mobile revolution," sending users into the real world armed with cameras, computers, gyroscopes, and microphones packed inside their smart new phones. In the competition for scope, surveillance capitalists want your home and what you say and do within its walls. They want your car, your medical conditions, and the shows you stream; your location as well as all the streets and buildings in your path and all the behavior of all the people in your city. They want your voice and what you eat and what you buy; your children's playtime and their schooling; your brain waves and your bloodstream. Nothing is exempt.

Unequal knowledge about us produces unequal power over us, and so epistemic inequality widens to include the distance between what we can do and what can be done to us. Data scientists describe this as the shift from monitoring to actuation, in which a critical mass of knowledge about a machine system enables the remote control of that system. Now people have become targets for remote control, as surveillance capitalists have discovered that the most predictive data come from intervening in behavior to tune, herd, and modify action in the direction of commercial objectives. This third imperative, "economies of action," has become an arena of intense experimentation. "We are learning how to write the music," one scientist said, "and then we let the music make them dance."

This new power "to make them dance" does not employ soldiers to threaten terror and murder. It arrives carrying a cappuccino, not a gun. It is a new "instrumentarian" power that works its will through the medium of ubiquitous digital instrumentation to manipulate subliminal cues, psychologically target communications, impose default-choice architectures, trigger social comparison dynamics, and levy rewards and punishments—all of it aimed at remotely tuning, herding, and modifying human behavior in the direction of profitable outcomes and always engineered to preserve users' ignorance.

We saw predictive knowledge morphing into instrumentarian power in Facebook's contagion experiments published in 2012 and 2014, when it planted subliminal cues and manipulated social comparisons on its pages, first to influence users to vote in midterm elections and later to make people feel sadder or happier. Facebook researchers celebrated the success of these experiments and noted two key findings: that it was possible to manipulate online cues to influence real-world behavior and feelings, and that this could be accomplished while successfully bypassing users' awareness.

In 2016, the Google-incubated augmented reality game *Pokémon Go* tested economies of action on the streets. Game players did not know that they were pawns in the real game of behavior modification for profit, as the rewards and punishments of hunting imaginary creatures were used to herd people to the McDonald's, Starbucks, and local pizza joints that were paying the company for "footfall," in exactly the same way that online advertisers pay for "click through" to their websites.

In 2017, a leaked Facebook document acquired by the *Australian* exposed the corporation's interest in applying "psychological insights" from "internal Facebook data" to modify user behavior. The targets were 6.4 million young Australians and New Zealanders. "By monitoring posts, pictures, interactions, and internet activity in real time," the executives wrote, "Facebook can work out when young people feel 'stressed,' 'defeated,' 'overwhelmed,' 'anxious,' 'nervous,' 'stupid,' 'silly,' 'useless,' and a 'failure.'" This depth of information, they explained, allows Facebook to pinpoint the time frame during which a young person needs a "confidence boost" and is most vulnerable to a specific configuration of subliminal cues and triggers. The data are then used to match each emotional phase with appropriate ad messaging for the maximum probability of guaranteed sales.

Facebook denied these practices, though a former product manager accused the company of "lying through its teeth." The fact is that in the absence of corporate transparency and democratic oversight, epistemic inequality rules. *They know. They decide who knows. They decide who decides.*

The public's intolerable knowledge disadvantage is deepened by surveillance capitalists' perfection of mass communications as gaslighting. Two examples are illustrative. On April 30, 2019, Mark Zuckerberg made a dramatic announcement at the company's annual developer conference, declaring, "The future is private." A few weeks later, a Facebook litigator appeared before a federal district judge in California to thwart a user lawsuit over privacy invasion, arguing that the very act of using Facebook negates any reasonable expectation of privacy "as a matter of law." In May 2019, Sundar Pichai, chief executive of Google, wrote in the *Times* of his corporation's commitment to the principle that "privacy cannot be a luxury good." Five months later, Google contractors were found offering five-dollar gift cards to homeless people of color in an Atlanta park in return for a facial scan.

Facebook's denial invites more scrutiny in light of another company document leaked in 2018 that offers rare insight into Facebook's computational factory, where a "prediction engine" runs on a machine intelligence platform that "ingests trillions of data points every day, trains thousands of models," and produces "more than six million predictions per second." But to what purpose? The report makes clear that these extraordinary capabilities are dedicated to meeting its corporate customers' "core business challenges" with procedures that link prediction, microtargeting, intervention, and behavior modification. For example, a Facebook service called "loyalty prediction" is touted for its ability to plumb proprietary behavioral surplus to predict individuals who are "at risk" of shifting their brand allegiance and alerting advertisers to intervene promptly with targeted messages designed to stabilize loyalty just in time to alter the course of the future.

That year a young man named Christopher Wylie turned whistleblower on his former employer, a political consultancy known as Cambridge Analytica. "We exploited Facebook to harvest millions of people's profiles," Wylie admitted, "and built models to exploit what we knew about them and target their inner demons." Mr. Wylie characterized those techniques as "information warfare," correctly assessing that such shadow wars are built on asymmetries of knowledge and the power it affords. Less clear to the public or lawmakers was that the political firm's strategies of secret invasion and conquest employed surveillance capitalism's standard operating procedures to which billions of innocent "users" are routinely subjected each day. Mr. Wylie described this mirroring process, as he followed a trail that was already cut and marked. Cambridge Analytica's real innovation was to pivot the whole undertaking from commercial to political objectives.

In other words, Cambridge Analytica was the parasite, and surveillance capitalism was the host. Thanks to its epistemic dominance, surveillance capitalism provided the behavioral data that exposed the targets for assault. Its methods of behavioral microtargeting and behavioral modification became the weapons. And it was surveillance capitalism's lack of accountability for content on its platform afforded by Section 230 that provided the opportunity for the stealth attacks designed to trigger the inner demons of unsuspecting citizens.

It's not just that epistemic inequality leaves us utterly vulnerable to the attacks of actors like Cambridge Analytica. The larger and more disturbing point is that surveillance capitalism has turned epistemic inequality into a defining condition of our societies, normalizing information warfare as a chronic feature of our daily reality prosecuted by the very corporations upon which we depend for effective social participation. They have the knowledge, the machines, the science and the scientists, the secrets and the lies. All privacy now rests with them, leaving us with few means of defense from these marauding data invaders. Without law, we scramble to hide in our own lives, while our children debate encryption strategies around the dinner table and students wear masks to public protests as protection from facial recognition systems built with our family photos.

In the absence of new declarations of epistemic rights and legislation, surveillance capitalism threatens to remake society as it unmakes democracy. From below, it undermines human agency, usurping privacy, diminishing autonomy, and depriving individuals of the right to combat. From above, epistemic inequality and injustice are fundamentally incompatible with the aspirations of a democratic people.

We know that surveillance capitalists work in the shadows, but what they do there and the knowledge they accrue are unknown to us. They have the means to know everything about us, but we can know little about them. Their knowledge of us is not for us. Instead, our futures are sold for others' profits. Since that Federal Trade Commission meeting in 1997, the line was never drawn, and people did become chattel for commerce. Another destructive delusion is that this outcome was inevitable—an unavoidable consequence of convenience-enhancing digital technologies. The truth is that surveillance capitalism hijacked the digital medium. There was nothing inevitable about it.

Existing privacy and antitrust laws are vital but neither will be wholly adequate to the new challenges of reversing epistemic inequality. These contests of the twenty-first century demand a framework of epistemic rights enshrined in law and subject to democratic governance. Such rights would interrupt data supply chains by safeguarding the boundaries of human experience before they come under assault from the forces of datafication. The choice to turn any aspect of one's life into data must belong to individuals by virtue of their rights in a democratic society. This means, for example, that companies cannot claim the right to your face, or use your face as free raw material for analysis, or own and sell any computational products that derive from your face.

On the demand side, we can outlaw human futures markets and thus eliminate the financial incentives that sustain the surveillance dividend. This is not a radical prospect. For example, societies outlaw markets that trade in human organs, babies, and slaves. In each case, we recognize that such markets are both morally repugnant and produce predictably violent consequences. Human futures markets can be shown to produce equally predictable outcomes that challenge human freedom and undermine democracy. Like subprime mortgages and fossil fuel investments, surveillance assets will become the new toxic assets.

In support of a new competitive landscape, lawmakers will need to champion new forms of collective action, just as nearly a century ago legal protections for the rights to organize, to strike, and to bargain collectively united lawmakers and workers in curbing the powers of monopoly capitalists. Lawmakers must seek alliances with citizens who are deeply concerned over the unchecked power of the surveillance capitalists and with workers who seek fair wages and reasonable security in defiance of the precarious employment conditions that define the surveillance economy.

Anything made by humans can be unmade by humans. Surveillance capitalism is young, barely twenty years in the making, but democracy is old, rooted in generations of hope and contest.

Surveillance capitalists are rich and powerful, but they are not invulnerable. They have an Achilles' heel: fear. They fear lawmakers who do not fear them. They fear citizens who demand a new road forward as they insist on new answers to old questions: *Who will know? Who will decide who knows? Who will decide who decides? Who will write the music, and who will dance?*

From the New York Times.

When Data Drives Decisions

MICHAEL BLOOMBERG

Michael R. Bloomberg is the founder of Bloomberg L.P. and Bloomberg Philanthropies, served three terms as mayor of New York City (from 2002 to 2013), and ran as a candidate for president in the 2020 Democratic primaries.

"In God we trust. Everyone else, bring data."

That was the motto I brought with me to government after spending my career in the private sector. The company I founded at the dawn of the computer age, Bloomberg L.P., used technology to put basic, real-time financial data at people's fingertips for the first time—helping them make better decisions about buying and selling stocks, bonds, and other financial instruments. In the market, data is power. But in government, data is often an afterthought, buried beneath political considerations, ideological attachments, and an overreliance on gut instinct. The result, sadly, is what people often expect from government: poor decisions, wasted resources, and ineffective problem-solving.

Today, a data-driven approach is sorely missing in Washington. During the Trump administration, those in charge showed no interest in facts and data. Can you imagine a manager, in any other line of work, presenting "alternative facts" with a straight face? Or dismissing the vast and overwhelming body of evidence that *human activity is changing the climate*?

In the real world, managers who present fiction as fact, or ignore basic reality, get fired.

But even if a government follows facts and respects data, that's only half the battle. Elected executives will find themselves unprepared to respond effectively to a crisis—whether pandemic, recession, or natural disaster—if they fail to prepare. Preparation starts with building strong teams, prioritizing the gathering and analyzing of data, listening to experts, and building partnerships with leaders in the nonprofit and private sectors. On public health, President Trump failed on each front, and those failures went on full public display when the coronavirus crisis engulfed the country and world.

Data is so essential to preparation not only because it leads to clarity but because it fosters creativity and innovation—and urgency. That process only happens if elected leaders demand it, hire agency leaders who can lead it, and refuse to settle for the same old fist-shaking. In New York City, we saw the power of technology and data to pinpoint problems, target resources, spur creativity, achieve results, and win over skeptics. We did that across every area we worked in, and here are just a few examples.

Public Spaces

"Are you out of your mind?!" That was my gut reaction as mayor of New York City in 2009 when our transportation commissioner, Janette Sadik-Khan, came to me with the idea of closing Times Square to auto traffic. Times Square—known as "The Crossroads of the World"—is one of the busiest intersections on the planet. Closing it to cars would be a recipe for gridlock! Or so it seemed at first thought.

But Sadik-Khan, working with the city's traffic engineers, had a counterintuitive idea: Closing Times Square to cars could actually *improve* traffic flow, as well as the neighborhood's overall quality of life. Could it work? We didn't know. But they had some preliminary data suggesting it could. Since traffic in the area was already terrible, I was willing to give it a try.

When we announced a six-month pilot project, the tabloids went crazy. Late-night comedians mocked it. Cab drivers said it would be a disaster. But we stuck with it—and we were able to evaluate the results because of a fortuitous change we had recently introduced: requiring all thirteen thousand yellow taxi cabs to include GPS devices. That enabled us to compare the flow of traffic before and after the closure—a perfect experiment. Sure enough, Janette and team were right: Overall traffic moved 7 percent faster after the closure than before it.

The critics could still complain, but they could no longer say—at least not honestly—that the closure had slowed traffic. The data proved otherwise. What was left to argue about? We gave 2.5 acres of public space back to pedestrians, and people loved it. The number of pedestrians injured in car crashes in the area fell 35 percent. The business community also approved. They now had even more customers walking by and through their doors each day. So we made the change permanent and began creating similar pedestrian plazas in other parts of the city. Other cities around the world took notice and followed suit. Instead of dismissing a counterintuitive, controversial idea, we rigorously tested our prototype and scaled it up only after it proved successful.

Air Pollution

When we set out to find the largest sources of the city's air pollution, we followed another tried-and-true management maxim: If you can't measure it, you can't manage it. The conventional wisdom was that cars and trucks accounted for most of the city's air pollution. Anyone who's ever sat in rush-hour traffic might have agreed. If we had worked under that assumption, we would have completely failed to clean the air and reduce the city's carbon footprint. Instead, we used technology that put our assumption to the test by installing air-quality monitors at street level. The results surprised us: Just 1 percent of the city's buildings were spewing more soot into the air than all of the city's cars and trucks combined.

Once we had that data, we developed a program specifically to help building owners convert to cleaner, more energy-efficient fuels, and we banned the dirtiest fuels. The building owners weren't happy about it, but we had the data to show them that their buildings were poisoning the air in their neighborhoods, and that made arguing against our position a lot tougher. Once again, data was power.

Combined, all our greening and sustainability initiatives helped shrink the city's carbon footprint by 13 percent. We knew that because we had done an inventory of greenhouse gas emissions before we started our work. The data showing the progress we were making on climate change helped win over skeptics. For those who were climate deniers, we could point to the improvement in air quality, which was cleaner than it had been in half a century. Who could argue with that?

That experience informed my work fighting climate pollution through my philanthropy. In America, cities account for 70 percent of greenhouse gas emissions. Cities are where the problem is, but they are also where the solution lies. Like New York, cities can reduce emissions in the areas where they typically exert the most control—buildings and transportation. My foundation has worked in partnership with mayors nationwide to spur those data-driven climate actions. Through a competition we launched called the American Cities Climate Challenge, we selected twenty-five winning US cities, and we are providing each with support to implement proven actions to reduce emissions.

Public-Private Partnerships

Government doesn't have all the answers. And rather than go boldly in search of answers, elected officials tend to be risk-averse. That, in a way, is understandable. Governments are responsible for spending taxpayer dollars wisely, and when budgets are tight, it can be hard to justify spending public money on untested ideas.

That's where private money comes in, which can fund ideas that, if successful, can have major public benefits. Private companies and philanthropists have the resources that governments need to innovate, and governments have the power to bring solutions to scale.

Partnerships on climate action are a perfect example. Climate change is an all-hands-on-deck emergency, and one effective way for government to respond is to work hand-in-hand with private partners. For example, California is teaming up on a new initiative with my foundation and an earth-imaging company called Planet. Our goal is to use satellite technology to inform and accelerate climate protection. Using existing data, we can analyze the operations of dirty coal-fired power plants and give the public a tool to see the worst polluters and hold them accountable. The next generation of satellites will be able to measure deforestation and detect greenhouse gases such as methane and carbon dioxide—the first step in addressing them faster and more effectively. Philanthropy can't replace government spending, but it can supplement it, help governments form these kinds of innovative partnerships, and innovate in ways that they can't with tax dollars alone.

It isn't enough for governments to collect data and sit on it, either. The real value is in how they use it—and here, again, is where the public sector can lean on the private sector. In New York City, we started the NYC BigApps competition, in which city government made a treasure trove of raw data (stripped of any personal information) available to software developers and the general public. In turn, a community of creative entrepreneurs got the chance to build mobile applications that not only improve New Yorkers' quality of life but also attract private investment. One success story is MyCityWay, which built an app that bundled city resources and sorted them by the user's location. MyCityWay won awards at the first BigApps competition, then it grew to include employees abroad and serve as a technology partner for leading global companies. When government brings data, and it's combined with private-sector talent and technology, we open up a world of possibilities.

Even as tech advances by leaps and bounds, however, the data it produces isn't always clear. Some leaders prefer to see only data that "their side" cherry-picks. Two people looking at the same data can arrive at two different conclusions. The old saying about there being three kinds of lies—"lies, damned lies, and statistics"—is especially true in politics. Anyone can find data to support their own positions, but using data honestly and effectively in government requires a rare level of nonpartisanship. Good data will never replace judgment. Data comes with no guarantees, but it can provide a common baseline for discussion and drive innovation and progress on even the most pressing challenges. Especially during a crisis, such as the coronavirus pandemic, it is absolutely essential we demand that elected officials respect data, communicate it honestly, and utilize it effectively—for the good of all people.

Tech, Heal Thyself

ELLEN PAO

Ellen Pao is one of Silicon Valley's leading advocates for fairness and ethics and is a longtime entrepreneur and tech investor. Her landmark gender discrimination case against venture capital firm Kleiner Perkins sparked other women, especially women of color, to fight harassment and discrimination in what's been called the "Pao effect."

We were dreaming of greenfields and imagining the sky's the limit when I started in tech in 1998. We aspired to revolutionize, reinvent, and reimagine the world. We strived to rebuild it from scratch, convinced we could optimize the future if we jettisoned the past. But the idealists who said they wanted to improve the world failed. We thought we were smarter than anything that already existed, but we should have started from a basic value already core to medicine: Do no harm.

Tech was touted as a savior but lost its way due to greed and insularity. How do we get back to wanting tech to be a benefit and actually making it one? How do we stop harming people and allow everyone to benefit? Diversity and inclusion is a path to save tech and could have prevented a lot of problems. If tech had been more diverse in contributors and if people had listened, we would not be stuck here today. I believe people would have seen these problems in advance—and avoided them.

In the early days, tech's potential seemed glorious, positive, meritocratic, democratic, and unlimited. Silicon Valley was pitched as uniquely "pure"—almost altruistic. It was a counter to Wall Street, which back then was the high-powered job goal for new MBAs: It was a fast track to generational wealth, prestige, power, and influence. The bad people joined Wall Street; the good people joined tech. We talked about technology's power to democratize by giving everyone access to everything: information, education, communication, you name it. It would connect all people to all other people for whatever reason they wanted.

The biggest example was Google, whose initial goal seemed clear and virtuous: "Organize the world's information and make it universally accessible and useful." Its better-known corollary was "don't be evil." It had lofty goals, like digitizing all the books ever written. Apple is another leading example; it launched the iPhone in 2007 as "the internet in your pocket." We lauded them for their ambitious goals and moon shots.

The situation, and our perception of it, started to get cloudier and uglier a decade later. Over on Wall Street, investment bankers' ability to generate wealth seemed to dissipate overnight. After the stock market dropped in 2008, greed turned into layoffs, bad press (vampire squids!), and for some, big fines and even jail terms. MBAs and other ambitious people redirected their paths to Silicon Valley as an expressway to riches—far beyond any wealth that an investment bank or hedge fund would generate, in less time, without having to have a boss, and without the risk of going to prison. In 2008, Google dropped to a $200 billion market cap; a year later, it had recovered its losses and was worth almost $400 billion. It generated four billionaires in less than a decade. Bill Gates was the richest person in the world for two decades.

To me, Facebook is the biggest example of, and possibly the largest contributor to, greed taking over. As would-be investment bankers looked for another road to riches, they say Mark Zuckerberg become a paper billionaire in less than five years. Its motto seems ominous today, "Move fast and break things." Microsoft's mantra when I worked there in 1998 was the slightly more innocuous "Ask for forgiveness, not for permission."

The third factor in this perfect storm was the corps of venture capitalists, who powered and accelerated this change. I had a front-row seat for seven years as an investing partner at a top VC firm; I saw firsthand how venture capitalists work—and how they fail to measure fundamental decency or hold people accountable. With the rapid success of Facebook, VCs intentionally changed what they were looking for and funding to single-mindedly focus on finding "the next Mark Zuckerberg." We funded the ability to sell a vision, not skills or experience in real-life execution. And it worked for a time. Great salespeople were able to keep raising money to move to the next step: more employees and more new users (no one cared how much they were subsidized). It was always a certain type of great salesperson. As VC titan John Doerr, who led Kleiner Perkins while I was there, told a room of venture capitalists in 2008, successful entrepreneurs "all seem to be white, male nerds who've dropped out of Harvard or Stanford, and they absolutely have no social life. So when I see that pattern coming in . . . it was very easy to decide to invest." He laid a clear path, and others followed.

If You Build It, They Will Come

Buzzwords like "network effect," "social proof," and "thought leader" had bigger impacts than "business model" or "unit economics." The widespread VC philosophy was that scale would provide the answer and solve all the problems. We moved from values to greed. Someone wrote about how it just takes one hot investment to make a VC fund successful, so you should pay up to get in. And VC firms did, driving up valuations across the board.

Instead of looking at ways to change communication, connection, and information flow, we began funding startups that aimed to replace our parents by doing laundry, buying groceries, restocking kitchens, delivering meals, and cleaning apartments. We then funded startups providing the millionaire life with on-call drivers, on-demand security guards, gourmet chefs, and personal assistants. The next trendy investments were startups that made it unnecessary to talk to a human being. We came up with Siri/Alexa/Google voice, startups for telemedicine and online therapy, automated drivers, and home drug delivery. It all became less about connecting people and more about replacing them with technology. These startups, and even the loftier ones before them, haven't been meritocratic, innovative, or at a baseline, ethical. Even the positive benefits cannot make up for the harms they caused. People have been left behind. People are being hurt. Too many are losing overall, not gaining, with technology.

In the rush to innovate, we forgot to pay attention to fairness. Much of Silicon Valley lore centers on eight white men who founded Fairchild Semiconductor in 1957 and went on to build and fund the tech industry. They hired their friends or people in their comfort zones, who were almost all white men. As the tech world grew, the founders became wealthy, and their employees did, too. They founded more startups, and some became investors who funded their friends or people in their comfort zones. And an industry centered on white men started to scale.

Today, the numbers tell the story. While we were told that tech is a meritocracy, the data show that it isn't, not by a long shot. In 2013, Tracy Chou started a project to count women engineers in tech startups. She crowdsourced data from tech companies and pushed them to share diversity reports and demographics. Her numbers for 2013 showed that women accounted for barely 18 percent of engineers at almost one hundred startups.

After that, companies started sharing more data. VC firms started sharing with *The Information*'s VC diversity index in 2015. Its 2018 report showed that the percentage of Black and Latinx decision makers in the US venture capital industry actually fell in 2018. According to Richard Kerby's research in 2018, 82 percent of venture capitalists are male and 70 percent are white. Less than 3 percent are Black, and only two-thirds of a percent are black women. Less than 2 percent are Latinx, and less than half a percent are Latinx women. Also in 2020, All Raise reported that 65 percent of US VC firms have no women partners, and women have only 13 percent of VC leadership roles, barely up from 12 percent in 2019.

The funding numbers are even more dire. In 2017, according to a 2019 RateMyInvestor report, only 1 percent of venture-backed founders were Black, only 1.8 percent were Latinx, and only 9.2 percent of founders were women. Eighty-two percent of founding teams were all-male. ProjectDiane 2018 reported that less than 4 percent of women-led startups were led by Black women and less than 2 percent were led by Latinx women.

The percentage of women—and of women of color from underrepresented groups—is far below representative in entrepreneur fundraising, in tech leadership, at VC firms and the boards they fill, and in the tech workforce. As a result, the percentages of wealth generated by women through equity is a sad 6 percent of the total equity from startups. The cycle continues.

None of the VC firms cared for the first fifty years. It had been that way since the start. White men funded white men and hired white men. It was a self-fulfilling prophecy; as almost only white men were funded, the vast majority of successful entrepreneurs were white men, further "proving" the pattern and perpetuating the cycle. Those white men hired more white men, and here we are. They don't hold one another accountable because they believe the system is working. They haven't paid attention to the disparity in opportunity for others, much less to the harm their products cause to the consumers who use them or are abused by them.

When I sued leading venture firm Kleiner Perkins for sexual discrimination and retaliation in 2012, reactions ranged from skeptical to outraged to dead silence. People called me crazy and much worse. I lost the trial in 2015, but I am proud to be part of a wave of more and more people speaking up. Some shared their stories anonymously in a *New York Times* article. Some wrote first-person essays about their experiences. Others joined an ongoing cadence of litigation. All exposed the fact that venture capital is broken, and the numbers describing homogeneity have resulted in a culture of sexual harassment and discrimination.

The numbers on harassment in the tech industry are damning: 78 percent of female founders said they or someone they know has personally experienced sexual harassment in the workplace. In another survey of women founders, 22 percent said they had personally experienced sexual harassment during their startup career.

This culture permeates the startups that VCs fund. The nonprofit I run, Project Include, surveyed nearly two thousand tech startup employees, and a third said they had personally seen or experienced inappropriate office behavior. Even more troubling, less than 3 percent felt comfortable reporting it. I hear from people privately, and their experiences are even worse than the ones that have been shared publicly. And companies that enable harassment seem to allow all kinds of bias beyond gender.

Unsurprisingly, these problems have manifested on social platforms—most often in the form of online harassment. When the internet was created, we thought we had all the answers and were bringing good to the world. Social platform creators touted "connecting people" (Facebook) and "digital public squares" (Twitter). We now know it hasn't met those expectations. We see women, transgender people, and people of color being harassed and bullied off these forums, especially when they have more than one underrepresented identity. We see election tampering, fake news, and technology-enabled genocide. We see cross-platform, online harassment turning into offline, real-life mass shootings. Hate is gaining traction, powered by tech.

Today, CEOs and founders have changed their minds about their social platforms' initial approaches and even about their platforms. Evan Williams, cofounder of Twitter, has said, "I thought, once everybody could speak freely and exchange information and ideas, the world is automatically a better place. I was wrong about that." The founders of 8chan and 4chan sold their sites, and the former publicly disavowed his, saying, "I sometimes wonder whether creating 8chan was a good thing. . . . Sometimes I think I should have been harder on violent threats." Dick Costolo, during his time as Twitter's CEO, wrote in an internal memo, "We suck at dealing with abuse and trolls on the platform, and we've sucked at it for years. . . . There's no excuse for it."

At reddit, we waited until 2015 to attack harassment on the site. When my team banned unauthorized nude photos and revenge porn, we were the first major platform to do so, and everyone else followed—but it was a longtime problem on all our sites. Now we see that harassment has moved from targeting female engineers starting in 2014 with the Gamergate online hate campaign to female leaders at reddit and female reporters, entertainers, politicians, and athletes. Women of color get even more abuse.

All this results in less participation from underrepresented groups; we've seen Lindy West, Ta-Nehisi Coates, and Kelly Marie Tran leave Twitter. In November 2019, in the United Kingdom, several female members of Parliament announced they were not running for reelection because of online harassment and threats. So much for the digital public square giving everyone a voice.

We see tech companies leaving many groups behind, including employees and contractors. Today, employees are speaking up and walking out over issues of employee equity. As wages drop and startups scramble to find profit in their heavily subsidized models, workers at the low end keep seeing their wages drop. Some are protesting the inequities in tech. Uber and Lyft drivers went on strike in nine cities in March and May 2019. Instacart shoppers went offline for three days in November 2019, with DoorDash and Postmates employees echoing their demands. It's no surprise that the last wave of pre-COVID-19 startups were helping people survive on inadequate income: rent my couch, rent my apartment, rent my car, rent my garage, rent my yard, rent my clothes. Wave after wave of Amazon employees protested working conditions in warehouses hit by COVID-19 infections. In March 2020, Amazon and Instacart employees went on strike at the same time, seeking better wages, sick leave, and safety protections in response to the COVID-19 pandemic. That same month, Facebook made headlines for allowing employees to work from home, but requiring contract workers to show up at the office.

Others are calling out a lack of ethics at tech companies, notably when they sell technology to help the government. Salesforce, Microsoft, Google, Palantir, and Amazon employees have objected to their companies providing software and services for the US Immigration and Customs Enforcement agency (ICE). Two speakers quit Microsoft's GitHub on the eve of its November 2019 global conference to protest. Google and Facebook faced broad criticism when they considered using their customer data to help governments measure whether social distancing policies were being followed during the COVID-19 pandemic.

We've learned that most tech leaders aren't listening. They aren't thinking about the people being left behind, even when their own employees are living in poverty and working in substandard conditions. Exposé after exposé has described horrendous working conditions for Facebook content moderators, Amazon deliverers, Uber and Lyft drivers, and Uber engineers. Lawsuit after lawsuit has attested to racism, sexism, and hate piled on employees at Tesla, Betterworks, Uber, Facebook, and more. The COVID-19 pandemic sharpened focus on disparities in the tech workforce as company after company failed to provide protective clothing, sick leave, or hazard pay for employees critical to getting necessities to customers sheltering in place safely at home.

But wealth at the very top is growing. Tech CEOs continue to dominate the billionaire listings. Jeff Bezos has the highest net worth, at over $100 billion. It's hard for them to acknowledge the problems, much less do the work to solve them. Again and again, we hear in the tech community that tech leadership will only react when public press calls attention to a problem, and often they respond with platitudinal apologies and vague, usually meaningless promises to fix things, or with a tax-favorable donation measured in basis points of their fortunes.

Is this really the best that tech can do? Obviously not. We need to go back to basics and rethink our values. We can and should focus on the principle of "do no harm." That means: Consider all costs. Create lasting values. Measure and hold ourselves accountable for harm to employees, partners, consumers, the environment, neighbors, and the public community.

Most importantly, bring in diversity at the highest levels. Research shows that diversity ultimately results in better decisions and creates more value. CEOs have acknowledged that diversity and inclusion improves the culture for everyone. The urgency to change is accelerating because VCs, tech leaders, and tech companies are now getting more public and government scrutiny.

Why can't our ambitious goals and moon shots include fair and equitable companies? What would that look like? How do we include all our employees, partners, and communities? How do we bring that go-big-or-go-home attitude to inclusion and accountability?

We can. We just need to hire people from all genders and races at the top, from the boardroom to executive offices. Only then will we make sure that people who use tech products will be paid fairly in wages and equity. We need to eliminate NDAs and forced arbitration so people can continue to litigate and share the stories that are catalyzing change. We need to look hard at the technologies that are increasingly invading our privacy, and the algorithms behind them, to avoid making the same mistakes in this next wave of innovation. Only then can we be sure that we are doing our best to do no harm.

We Need a New Capitalism, Based on Trust

MARC BENIOFF

Marc Benioff is chairman, chief executive officer, and founder of Salesforce and a pioneer of cloud computing.

At the annual World Economic Forum in early 2018, the tech industry experienced an acute crisis of trust. Packed into the giant auditorium, in every seat and lining the walls, was a veritable army of global influence: leading politicians, executives, bureaucrats, academics, journalists, and policy experts. On either side of me sat my four fellow panelists, an esteemed group of technology CEOs and thought leaders. Behind us, the title of the keynote panel that was about to commence flashed on the screen in giant, intimidating type. It wasn't a title, really—more like a loaded question: *In Technology We Trust?*

As we sipped water and smoothed the creases in our jackets, our moderator, the business journalist Andrew Ross Sorkin, kicked things off by declaring that the public's eroding confidence in the judgment of technology companies was "the most pivotal and important discussion that's taking place right now, really in industry anywhere." He was right.

Recent revelations of alarming privacy breaches at Facebook, and even possible Russian-backed voter manipulation in the 2016 presidential election, were not sitting well with a large portion of the American public. The burning question on many of the minds gathered in Davos, Switzerland, was deceptively simple: What were we going to do about it?

Technology is a field to which I have devoted my entire career. I've often marveled at how today's advances in computing have transformed the way we live and work. During the last decade—the start of a period often referred to as the Fourth Industrial Revolution—we saw extraordinary advances in artificial intelligence, quantum computing, genetic engineering, robotics, and 5G connectivity. Massive tributaries of digital information are now flowing at a speed and scale unthinkable even a decade ago, while artificial intelligence (AI) and robotics are breaking down barriers between human beings and machines. Everyone and everything on the planet is becoming connected, creating complex business challenges and disruptions nobody could have foreseen.

I've always believed that technology has the potential to reshape our world in wonderful ways—to foster a more open, diverse, trusting, inclusive, and democratic society and to create once-unthinkable opportunities for billions of people. Today more people on the planet have mobile phones than running water or electricity in their homes. Billions of people have been lifted out of extreme poverty. Between 2000 and 2016, global average life expectancy increased by 5.5 years—the fastest increase since the 1960s.

But technology, clearly, is no panacea. New pressures and dangers have emerged, and with them moral conundrums that none of us ever considered. The global inequality gap has caused an erosion of trust in institutions, while complex social and economic issues—including privacy, ethics, education, the future of work, and the health of the planet—have begun to insert themselves, often uncomfortably, into the corporate agenda.

In the panel's opening minutes on that snowy morning in Davos, Ruth Porat, CFO of Alphabet (parent company of Google), remarked that she believed the public hadn't lost trust in Google; after all, people kept returning to its platforms daily to perform trillions of searches. It was a line of argument I'd heard from my tech colleagues many times before.

When it was my turn to speak, I said, "Trust has to be your highest value in your company. And if it's not, something bad is going to happen." I could sense a ripple of discomfort in the room. After a short pause, I pointed out other moments in history when regulators had fallen asleep at the wheel. I mentioned tech in the same breath as credit-default swaps, sugar, cigarettes—harmful products that companies had been allowed to peddle to customers, unconstrained by regulations. Our industry had been given a regulatory pass for years, I continued. "When the CEOs won't take responsibility," I said, "then I think you have no choice but for the government to come in." I shared my belief that Facebook is the "new cigarettes"—addictive and harmful to kids—and I called out others who focus on more users and profits and don't make values, especially trust, their highest priority.

When I arrived home from Davos, my phone began ringing nonstop. Leaders in the technology industry kept calling, one after another, to inform me that I had betrayed them. Apparently, I had broken ranks or crossed some imaginary line, and they didn't like it. My wife, Lynne, jokingly started calling me "the regulator."

Yet since that snowy weekend in 2018, more voices have joined mine, wondering if checks and balances should be imposed on the effects of technology on societies and individuals. In 2019, the Business Roundtable, an association of CEOs of leading US companies, announced a new "Statement on the Purpose of a

Corporation," committing to lead their companies for the benefit of all stakeholders—customers, employees, suppliers, communities, and shareholders. As business leaders, government officials, educators, and citizens, we need to create a common set of principles and values that take us, together, to a future that we all want. And we need to act on those principles and values.

From the outset, I had hoped that Salesforce, the company I founded in 1999, would prosper by traditional measures. But I was equally determined that it would have a positive impact on the world. We identified the values that would serve as our bedrock: trust, customer success, innovation, and equality. We also decided that no matter how big the company grew, we would always set aside 1 percent of our equity, product, and employee time for charitable causes—an initiative we call the 1-1-1 philanthropic model, now adopted by more than nine thousand companies from over one hundred countries through Pledge1percent.org.

In the twenty-one years since founding Salesforce, I've become convinced that there are two types of CEOs: those who believe that improving the state of the world is part of their mission, and those who don't feel they have any responsibility other than delivering results for their shareholders.

This schism is more pronounced now that the world is faced with the sheer power of big tech and social networks as crucial players in our democracy, society, and even our private lives. Their innovation and influence have outpaced adequate scrutiny of their shortcomings in protecting privacy and stemming disinformation and hate speech. In the past, far more chief executives have clung to their fiduciary duty to shareholders, while their employees, communities, and the world at large took a back seat. I don't condone that instinct, but I understand it completely. Back when I was in business school in the 1980s, I studied the immortal words of the economist Milton Friedman: "There is one and only one social responsibility of business," he wrote in his book *Capitalism and Freedom*. The answer, of course, was to "increase its profits." In an essay for the *New York Times* in 1970, Friedman went so far as to argue that executives who claim that companies have "responsibilities for providing employment, eliminating discrimination, avoiding pollution, and whatever else may be the catchwords" of the day are guilty of "undermining the basis of a free society."

With all due respect, Milton Friedman was wrong. He was wrong then, and he's doubly wrong in today's context. The business of business isn't just about creating profits for shareholders. We're simply too big, too global, and too immersed in people's daily lives. Yes, our business is to increase profits, but our business is also to improve the state of the world and drive *stakeholder*—as well as shareholder—value. Not just because serving the interests of all stakeholders is good for the soul; it's good for business.

Statistics bear this out. The 2018 Deloitte Millennial Survey found that millennials believe business success should be measured by more than profits, citing, as their top priorities, the creation of innovative ideas, products, and services; positive impact on the environment and society; job creation, career development, and improving people's lives; and promotion of inclusion and diversity in the workplace. According to the 2019 Edelman Trust Barometer, 75 percent of consumers say they won't buy from unethical companies, and 86 percent say they're more loyal to ethical companies. Research from JUST Capital found that companies that do best in balancing the interests of all stakeholders generate a 6.4 percent higher return on equity and have a higher valuation from investors relative to their peers.

But corporate leadership seems to be lagging behind. LRN, a leading ethics and compliance company, surveyed eleven hundred executives, managers, and employees, and found that 87 percent agreed that the need for moral leadership in business is greater than ever. Yet only 7 percent of employees surveyed said their leaders often or always exhibited those behaviors. The disconnect between beliefs and action remains enormous, and there will be consequences for companies whose leadership doesn't live by values like trust and equality. In this age of instantaneous digital feedback, they and their leaders can no longer turn a blind eye to the issues that matter to their employees, their customers, and the communities in which they do business.

It's not a coincidence that in recent years more CEOs are starting to speak out on social and political issues—as a matter of survival, if nothing else. Let's face it, with government and other powerful institutions getting increasingly bogged down in political partisanship, brinkmanship, and perpetual gridlock, corporate participation is becoming more necessary. The deepening crisis of trust—as well as the growing educational class divide, income inequality, and the massive environmental challenges we face—make it impossible to abdicate responsibility and stay on the sidelines.

That's why companies have the potential to become the greatest platforms for change and create a future that benefits all stakeholders. Consider this one fact: 70 percent of the top one hundred revenue-generating entities in the world are corporations, not nations. I'm talking about Walmart, Apple, Samsung, and Exxon. The people who work at these kinds of companies, from the CEO down to the newest employee, have not only the responsibility but the resources, economic muscle, and implicit permission to be courageous on social issues and effect real change.

I've long been inspired by Unilever's recently retired CEO, Paul Polman, who made activism part of his corporate strategy while at the helm. He's been outspoken in advocating for companies to produce healthier products, improve worker conditions, and adopt renewable energy sources. "We have to bring this world back to sanity," he said, "and put the greater good ahead of self-interest." During his ten years as CEO, Unilever's stock more than doubled, proving again that customers spend their money with companies that uphold the things that they believe in, too.

When my new book, *Trailblazer: The Power of Business as the Greatest Platform for Change*, was published in fall of 2019, I called for a new capitalism that values all stakeholders—customers, employees, homeless, public schools, and our planet—as well as shareholders.

Yes, investors are a key stakeholder. Since becoming a public company in 2004, Salesforce has delivered a 3,500-percent return to our shareholders. But our communities aren't just our mailing addresses; they're key stakeholders, too. That's why we've given $330 million in grants to worthy causes, why our employees have volunteered five million hours, and why forty-six thousand nonprofits and NGOs use $1 billion of our software at no cost each year.

With this new capitalism, values create value. If trust isn't your highest value, your employees will walk out, your customers will walk out, and your investors will walk out. You will lose your jobs. We're seeing that more and more every day.

The old capitalism said you couldn't deliver both shareholder return and stakeholder return. That's a false choice. A 2018 article in the *Harvard Business Review* perhaps summed it up best, declaring that "CEO activism has entered the mainstream." According to the authors, Aaron K. Chatterji and Michael W. Toffel, this is just the opening wave of what has become the new norm. "The more CEOs speak up on social and political issues, the more they will be expected to do so," they write, adding, "in the Twitter age, silence is more conspicuous—and more consequential."

We used to think of customers, employees, and communities—local and global and everything in between—as different constituencies. But really, they aren't so different after all. They're all part of the larger ecosystem that our companies serve. And they're united in demanding that we deliver them the innovative products they want, and deliver on our commitment to uphold the values they care about.

The technology industry has a critical role to play, as the Fourth Industrial Revolution unleashes new transformative capabilities. Technology is neither good nor bad. Much like our words, it's just a tool; what really matters is how you use it. In the end, ensuring that it is used ethically is central to maintaining the world's democratic societies. We need regulations governing how AI interacts with humans and how digital agents protect privacy and enable individuals to control their own data, as well as regulations to remove biases within AI models and training data that perpetuate discrimination against various groups of people. We need to regulate technology platforms that are used to spread conspiracy theories and hate and to run knowingly false ads.

Put simply, tech companies can no longer wash their hands of what people do with our products.

At Salesforce, we established our own Office of Ethical and Humane Use, which works across product, law, policy, and ethics to develop and implement a strategic framework for the ethical and humane use of technology across the company. As issues arise, we don't want to rely on instinct or the political winds. Instead, we have a diverse group of expert advisers and a process to evaluate each situation. This office, for example, recently banned customers who sell the types of assault weapons used in mass shootings.

There's another critical issue arising from AI. Along with a new level of productivity and automation will come job dislocation. The shift will be as profound as the move from agrarian to manufacturing jobs. By one estimate, nearly half of all jobs worldwide could be impacted over the next two decades due to AI and automation. At the same time, entirely new categories of jobs are emerging to replace them. As a society, we need to adapt to the changing nature of work by focusing on training people for the jobs of tomorrow. By another estimate, 65 percent of children entering primary school today will have jobs in categories that don't yet exist. We have to start by addressing shortfalls in education systems from the earliest age.

The United States is the wealthiest nation in the world, yet more than 1.5 million American children live in poverty. In my community, the San Francisco Bay Area, we are facing a homelessness epidemic that is impacting the most vulnerable. The San Francisco Unified School District reports that one in twenty-six children in our public schools are homeless or marginally housed. Studies by the University of California, San Francisco, have shown that children who do not have the appropriate education and health care opportunities by age five will remain at a disadvantage for the rest of their lives.

We need to invest in giving every child the opportunity to get off to a good start and prepare them for jobs in a world where people and intelligent machines are working together. At Salesforce, for example, we started a partnership with our local school districts to help improve computer science education. Through this partnership, San Francisco became the first school district in the United States to establish a computer science curriculum for all grades. With additional full-time coaches and teachers for mathematics and technology instruction, class sizes for eighth-grade mathematics have been reduced by 25 percent, improving the learning environment for thousands of students.

It's natural for a large, growing company with a healthy commitment to building a better world to focus on the weightiest global issues. But I've found that one of the most meaningful acts of giving back for me, and for our fifty thousand employees, is volunteering at our local public schools. At Salesforce, every executive adopts a public school, and we give our employees seven paid days off each year to volunteer in schools (or wherever they choose). But children and teenagers are no longer the only demographic in need of better access to educational resources.

According to the World Economic Forum's 2018 *The Future of Jobs Report,* more than 50 percent of all employees will require some reskilling by 2022. People working in sales and manufacturing, for example, will need to acquire more technical skills. Today, at least one in four workers across all Organization for Economic Cooperation and Development (OECD) countries is already reporting a mismatch with regards to the skills demanded by their current job. Despite the growing need for adult reskilling and job training, such opportunities are not currently available or accessible for most people. Millions of people transitioned out of work by machines and algorithms would deal a giant blow to global and local economies.

I view this as a once-in-a-generation opportunity to address the inevitable economic dislocation that will result from technological innovation. This means ensuring an adequate form of job security for those people who find that their careers are disrupted or replaced. It's likely that more jobs will be created than lost in the coming decade, so investing in training and reskilling is a critical step in enabling displaced workers to reenter the workforce.

We need to do nothing short of reimagining the social contract for the twenty-first century—and a challenge this big and complex cannot be left to politicians alone. Rather than sitting on the sidelines as automation accelerates, every business has a vested interest in finding ways to train and reskill workers for the jobs of tomorrow.

In the United States there are more than 600,000 open tech jobs, but our universities produced only 78,000 computer science grads in 2019. Meanwhile, our companies can be incredible universities for educating the workforce of the future. This is why we invest in training employees, as well as interns and apprentices, so they can acquire new skills—in many cases through specialized instruction and hands-on experience that can't be obtained at even the most prestigious universities.

JPMorgan Chase, for example, is spending hundreds of millions of dollars funding community college and other nontraditional career-development programs globally for women, veterans, and underrepresented minorities. Dow, IBM, and Siemens have established apprenticeship programs to help fill the skills gaps in their industries.

At Salesforce we also use our free online learning platform, Trailhead, to help millions of people develop the new skills required in today's rapidly changing digital economy. With Vetforce, we offer free training to US service members, veterans, and spouses so they can acquire the skills necessary to obtain tech jobs. I was personally inspired by T. J. McElroy, who, after losing his sight in the Marine Corps, became a certified information-technology administrator through the Vetforce program. Today, he instructs other disabled vets to prepare them for technology careers.

Our companies, in fact, are vast armies of millions of people who could have a tremendous impact on reskilling workers: mentoring them, working side by side with them, and giving them the tools they need to learn, grow, and succeed in the digital age. By coming together—not just with other companies but with community colleges, universities, veterans' groups, NGOs, K–12 schools, and governments—we can help close the skills gap, nurture prospective employees, and develop the future workforce that will support a strong and growing economy.

In building tomorrow's workforce, we need to ensure that we achieve gender equality: equal pay for equal work. At Salesforce, we conducted a pay audit and found unexplained differences in pay between men and women. As a result, we have spent more than $9 million over the last few years to close the gap.

As the technologies of the Fourth Industrial Revolution burrow ever more deeply into our lives, we face what should be an easy choice. We can either use these advances to benefit all of humanity and preserve our environment for generations or benefit the few and continue to degrade our planet. For example, will we use AI-powered drones and satellites to expedite overfishing? Underwater robots to mine and pillage our ocean floor? Or will we use drones and satellites to monitor our oceans and stop illegal fishing, and use robots to track the salinity, temperature, and oxygen levels in our seawater?

We're at a crossroads, and the actions we take, from this moment forward, will have a defining impact on what kind of world we leave behind for future generations. The technology revolution we are in the midst of must be accompanied by a trust revolution.

I genuinely believe we are on the brink of a Fifth Industrial Revolution, one in which trust will be earned by those companies that apply the technologies being developed today to improve the world. In the future, innovation and democracy cannot advance in a positive direction unless they are grounded in genuine and continued efforts to lift up all of humanity. Companies, and the people who lead them, can no longer afford to separate business objectives from the social issues surrounding them. They can no longer view their mission as a set of binary choices: growing or giving back, making a profit or promoting the public good, or innovating or making the world a better place. Rather, it must be "and." Doing well by doing good is no longer just a competitive advantage. It's a business imperative.

As the CEO of a tech company, I speak daily to executives from businesses across every industry and part of the globe. From that vantage point, I have reason to look at our future with optimism. I believe every company and every individual—from new hires to those sitting in the corner office—has the potential to become a platform for change. A better future depends on all of us.

Excerpted from Trailblazer: The Power of Business as the Greatest Platform for Change *(New York: Currency, 2019).*

Most Tech Companies Don't Care

NICHOLAS D. KRISTOF AND SHERYL WuDUNN

Nicholas D. Kristof is a two-time Pulitzer Prize winner whose op-ed columns appear twice a week in the New York Times. *He is coauthor with his wife, Sheryl WuDunn, of five bestselling books. Sheryl WuDunn is the first Asian-American reporter to win a Pulitzer Prize and is cofounder of FullSky Partners, a consulting firm focused on socially driven ventures in technology and health care.*

When it comes to creating social good, Americans used to think that the polar opposites were greedy corporations and bleeding-heart nonprofits trying to save the world. In fact, there's a growing recognition that this divide is misleading and that for-profit companies sometimes do more good than charities. What matters is impact, not earnestness. A large company doing good has scale that is typically far greater than that of a nonprofit, and in the right situation it manages to: a) serve the public good; b) associate its brand with its good work; and c) gain in recruitment of millennials who care about making the world a better place.

Delta Air Lines, for instance, decided that it would address human trafficking: It has donated $2.5 million to sponsor a hotline run by Polaris Project, flown one hundred trafficking survivors to safe places in the past few years, offered apprenticeships to a few survivors, trained sixty-six thousand employees to spot victims, and created a video about trafficking that aired on seat-back screens. With 71 percent of labor trafficking victims reportedly transported into the United States on airplanes, the mission is a good fit, and it has been enthusiastically endorsed by the CEO, Ed Bastian. At modest cost, it has won Delta public praise, and the company says these efforts have helped morale and supported high employee retention rates.

So the puzzle is this: Why are technology companies so wretched at helping make the world a better place? Why do they seem so indifferent to social issues? Why are tech companies laggards in this area rather than leaders?

These behemoths depend on attracting millennial coders, so recruitment should be a top priority. Branding is paramount for companies like Apple, and public goodwill is critical in the long run for companies like Facebook or Amazon that are trying to stave off increased scrutiny and regulation. Yet the truth is that technology companies, which typically pride themselves on being ahead of the curve, have mostly been slow and weak in finding ways to give back and support local communities.

We were early enthusiasts of the potential of companies to make a difference. We wrote in an earlier book, *A Path Appears*, and argued in countless lectures, that what mattered was not an entity's tax status but whether it actually benefited humanity. We championed the idea that as companies embraced corporate social responsibility—or, even better, embraced the public good as a core part of the company mission—this would have a far greater impact on injustice and inequity than any number of foundations or charities. We were excited as companies and corporate groups talked about moving from "shareholder capitalism" to "stakeholder capitalism," with attention paid to the interests of employees and community members as well as stockholders. One of us, Sheryl, has taught seminars on impact investing and currently advises socially driven companies through FullSky Partners.

Yet we have been mostly aspirational about all this, at least so far. Companies talk a good game, but they haven't stepped up. Even when one branch of an industry has done good, other branches have behaved egregiously. Some pharmaceutical companies, for instance, despite their generous donations of medicines, have turned out to be engaged in "drug trafficking" that would make a Colombian drug lord blush, and they are defending lawsuits right now. Even McKinsey, once thought the gold standard for corporate America, has been enmeshed in scandal after scandal. While there is some movement in much of corporate America, and boards are showing greater interest and sometimes designating a board member to oversee social responsibility, technology companies are still mostly asleep at the wheel. They have certainly not been pioneers, as we might have expected.

Few companies depend more on brand-related goodwill than Apple, yet few big companies have done so little over the last decade. Facebook is facing increasingly hostile regulators worldwide. It is located right next to one of the most disadvantaged communities in America, East Palo Alto, yet it barely engages. The Amazon wilderness does a great deal for humanity, but the Amazon company? Not so much. And so on and so on.

It's true that Mark Zuckerberg and Priscilla Chan have been personally generous in their philanthropy. Ditto for Laurene Powell Jobs and Emerson Collective. And of course, the Bill and Melinda Gates Foundation has been the gold standard of philanthropy, not to mention the David and Lucile Packard Foundation, founded in 1964, and the Hewlett Foundation, launched in 1966. Individuals within the tech community have often been committed philanthropists and profoundly intelligent in the causes they support. They often are willing to take risks in ways that long-established foundations do not, and they bring an emphasis on metrics and results that is much needed. But while individuals have been generous, companies themselves mostly have not been—and we believe this is shortsighted for them, as well as a lost opportunity.

The two best examples of industries finding channels to create public goods are law firms (with pro bono work) and pharmaceutical companies (with pharma donations). They have created billions of dollars over the years in social value, defending the rights of the indigent and saving lives with donated medicines. These are not industries that are intuitively likely to step up. Lawyers are caricatured as greedy egoists, and pharma companies engaged in enormous wrongdoing during the opioid crisis. Yet major law firms donate large numbers of hours to good causes, and these are taken seriously: Helping a partner's granddaughter register her lemonade stand does not count. Large numbers of people have been freed from death row because of aggressive litigation by top-flight law firms. Major law schools have been cultivating the habit of performing pro bono work among students for years, promoting the concept that everyone deserves access to justice. The American Bar Association also boosts participation in pro bono programs at law firms by publishing reports—a mild form of shaming—on how well firms help "people with limited means." In 2018, some 130 of the biggest law firms devoted more than five million hours, a twenty-year record, to pro bono work. That amounted to 3.8 percent of billable hours, with the majority of time going toward helping low-income people gain access to legal services.

Likewise, pharmaceutical companies have donated vast numbers of medicines to deworm children, fight malaria, fight blindness, and so on. Merck's donations have vastly reduced the scourge of river blindness and made large parts of West Africa cultivatable again. This is not, of course, because pharma executives are particularly upstanding: This is the same sector that recklessly peddled opioids and killed thousands of Americans with their greed and irresponsibility. Yet whatever their sins, the corporate social responsibility was real and made an enormous difference. We believe that what sets these two industries apart is that there is rigorous outside monitoring of their corporate social responsibility, and this forces companies to actually compete in this space. *American Lawyer* magazine publishes data on pro bono work that is widely consulted in the recruitment process, and a similar index rates pharma companies on their drug donation programs. One of our reasons for writing this essay is the hope that outside scrutiny and prodding will poke some life into tech companies in this realm.

If tech companies want a model, they need only look at Unilever. It has been a leader in emphasizing social impact, with responsible sourcing and a focus on environment and sustainability. Paul Polman, the former CEO who publicly articulated Unilever's plan for sustainability, eliminated quarterly earnings reporting (though it releases sales and other results) as well as earnings guidance. Polman also said he wasn't interested in hedge funds as investors. Unilever has set ten-year plans for its social impact, with aggressive goals in the environment and society: reduce the amount of water used in manufacturing, reduce manufacturing waste, boost sustainable packaging and sustainable washing, reduce its carbon footprint, minimize health and safety risks throughout its supply chain, and eliminate human rights abuses. It is still a work in progress, and sales growth has been uneven in 2019, but its consumer image has strengthened over the years. Moreover, Unilever is doubling down on its bet. In 2019, its new CEO, Alan Jope, said Unilever might shed products, like Marmite food spread and Pot Noodle instant soups, that don't score high enough on the company's "purpose" meter.

Where are the technology companies? With their billions in cash reserves and sky-high margins, they have been high-profile absentees. There was some bold talk years ago about bridging the technology divide, but nothing much happened; to the extent that there was progress, it involved the US government, not tech companies, bringing bandwidth to rural communities. Indeed, there's a widespread perception that technology companies have created negative externalities in their communities by raising housing costs from Seattle to San Francisco in ways that are devastating to police officers, nurses, retirees, and others with more modest incomes who still need to find a place to live.

We suspect that part of the reason the technology sector lagged was hubris: Its executives were full of themselves, hyperfocused on the business, and rested on their laurels. When we spoke to tech company executives, they sometimes claimed that their products were public goods and fulfilled their public service obligations—corporate social responsibility was their daily bread and butter. Those were the days when technology executives walked on water and people talked about Mark Zuckerberg running for president.

Now that the mood has shifted, tech companies should see that this is not only a matter of doing right, but it is also in their interest. Boards are remiss in not elevating this issue, for they hurt their brands with their tardiness in addressing these concerns. These companies depend upon highly educated young people and many millennials, both for staffing and as influencers who buy or use their products. And this generation cares deeply about social justice. While baby boomers have grown accustomed to think that "doing good" means writing checks to charities at the end of the year, millennials take a far different approach. For them, doing good is about the company they work for, the brands they patronize, the stores they shop in, and the investments they make. It's not one hour in December, but 24/7. So far, tech companies have managed to sustain recruitment and retention despite a poor record on values; we believe that this will become more difficult.

Companies themselves seem to recognize this new reality, and slowly, grudgingly, we see some movement. Apple has scrutinized its supply chain and brought in third-party monitors to catch violations. It has also built a new environmentally friendly headquarters and tried to clean up environmental practices among suppliers in China. More recently, companies have been jockeying to show concern for affordable housing. In January 2019, Microsoft fired an opening salvo when it announced that it would donate $500 million to develop affordable housing and help ease the housing crisis in Seattle. In June, Google pledged $1 billion to help build affordable housing in San Francisco, where a shortage of some 3.4 million units persists. Then, in October 2019, Facebook committed $1 billion to the same cause. Not to be outdone, Apple announced in November that it was donating $2.4 billion to ease the housing crisis in partnership with the state of California.

This is a beginning, but a full embrace of doing good still seems to be outside the DNA of these companies. There will be a process of naming and shaming, of public badgering, that will force them to understand that elegant code is not enough for success and that they have to do better for their communities. They have to convey to employees and customers that they have values—not just Ping-Pong tables, free food, and nap rooms. Boards are supposed to supply outside scrutiny and adult supervision, so they should raise these matters in every meeting. The tech sector has enormous potential to make this a better world, but so far that potential has been largely unrealized.

An Open Letter to Mark Zuckerberg

AARON SORKIN

A screenwriter, director, producer, and playwright, Aaron Sorkin wrote and directed The Social Network—*a film adapted from Ben Mezrich's 2009 book* The Accidental Billionaires, *about Facebook's founding and the resulting lawsuits.*

October 31, 2019

Mark,

In 2010, I wrote *The Social Network* and I know you wish I hadn't. You protested that the film was inaccurate and that Hollywood didn't understand that some people build things just for the sake of building them. (We do understand that—we do it every day.)

I didn't push back on your public accusation that the movie was a lie because I'd had my say in the theaters, but you and I both know that the screenplay was vetted to within an inch of its life by a team of studio lawyers with one client and one goal: Don't get sued by Mark Zuckerberg.

It was hard not to feel the irony while I was reading excerpts from your recent speech at Georgetown University, in which you defended—on free speech grounds—Facebook's practice of posting demonstrably false ads from political candidates. I admire your deep belief in free speech. I get a lot of use out of the First Amendment. Most important, it's a bedrock of our democracy and it needs to be kept strong.

But this can't possibly be the outcome you and I want, to have crazy lies pumped into the water supply that corrupt the most important decisions we make together. Lies that have a very real and incredibly dangerous effect on our elections and our lives and our children's lives.

Don't say Larry Flynt. Not even Larry Flynt would say Larry Flynt. This isn't the same as pornography, which people don't rely upon for information. Last year, over 40 percent of Americans said they got news from Facebook. Of course the problem could be solved by those people going to a different news source, or you could decide to make Facebook a reliable source of public information.

The tagline on the artwork for *The Social Network* read, in 2010, "You don't get to 500 million friends without making a few enemies." That number sounds quaint just nine years later because one-third of the planet uses your website now.

And right now, on your website, is an ad claiming that Joe Biden gave the Ukrainian attorney general a billion dollars not to investigate his son. Every square inch of that is a lie and it's under your logo. That's not defending free speech, Mark, that's assaulting truth.

You and I want speech protections to make sure no one gets imprisoned or killed for saying or writing something unpopular, not to ensure that lies have unfettered access to the American electorate.

Even after the screenplay for *The Social Network* satisfied the standards of Sony's legal department, we sent the script—as promised over a handshake—to a group of senior lieutenants at your company and invited them to give notes. (I was asked if I would change the name of Harvard University to something else and if Facebook had to be called Facebook.)

After we'd shot the movie, we arranged a private screening of an early cut for your chief operating officer, Sheryl Sandberg.

Ms. Sandberg stood up in the middle of the screening, turned to the producers who were standing in the back of the room, and said, "How can you do this to a kid?" (You were twenty-six years old at the time, but all right, I get it.)

I hope your COO walks into your office, leans in (as she suggested we do in her bestselling book), and says, "How can we do this to tens of millions of kids? Are we really going to run an ad that claims Kamala Harris ran dog fights out of the basement of a pizza place while Elizabeth Warren destroyed evidence that climate change is a hoax and the deep state sold meth to Rashida Tlaib and Colin Kaepernick?"

The law hasn't been written yet—yet—that holds carriers of user-generated internet content responsible for the user-generated content they carry, just like movie studios, television networks, and book, magazine, and newspaper publishers. Ask Peter Thiel, who funded a series of lawsuits against Gawker, including an invasion of privacy suit that bankrupted the site and forced it to close down. (You should have Mr. Thiel's number in your phone because he was an early investor in Facebook.)

Most people don't have the resources to employ a battalion of fact-checkers. Nonetheless, while you were testifying before a congressional committee two weeks ago, Representative Alexandria Ocasio-Cortez asked you the following: "Do you see a potential problem here with a complete lack of fact-checking on political advertisements?" Then, when she pushed you further, asking you if Facebook would or would not take down lies, you answered, "Congresswoman, in most cases, in a democracy, I believe people should be able to see for themselves what politicians they may or may not vote for are saying and judge their character for themselves."

Now you tell me. If I'd known you felt that way, I'd have had the Winklevoss twins invent Facebook.

From the New York Times.

Technology and Social Connection

VIVEK MURTHY

American physician Vivek Murthy served as the nineteenth Surgeon General of the United States, from December 2014 to April 2017.

When I graduated from medical school and embarked on my training to become an internal medicine physician, I expected that most of my time would be spent caring for people with obesity, heart disease, mental illness, diabetes, and addiction to tobacco and other drugs. Years later, when I began my tenure as the nineteenth Surgeon General of the United States, I expected these public health epidemics to again be at the center of my work. As I visited communities across the country in my role as the "nation's doctor," many of these health concerns did in fact come up in conversation.

But one recurring topic was different. It wasn't a frontline complaint. It wasn't even identified directly as a health ailment. It was loneliness, and it ran like a dark thread through many of the more obvious issues that people brought to my attention—like addiction, violence, anxiety, and depression. Teachers, school administrators, and many parents I encountered, for example, voiced a growing concern that our children were becoming isolated—perhaps, especially, those who spent much of their time in front of their digital devices and on social media.

What Is Loneliness?

Loneliness is the subjective feeling that you're lacking the social connections you need. It can feel like being stranded, abandoned, or cut off from the people with whom you belong, even if you're surrounded by other people. What's missing when you're lonely is the feeling of closeness, trust, and the affection of genuine friends, loved ones, and community.

Unlike the subjective feeling of loneliness, isolation describes the objective physical state of being alone and out of touch with other people. But being physically alone doesn't necessarily translate into the emotional experience of loneliness. We can feel lonely even when we're surrounded by other people.

As I would later realize, the loneliness I was seeing and hearing about around the country was far more common than I thought. According to a 2018 report by the Henry J. Kaiser Family Foundation, 22 percent of all adults in the United States say they often or always feel lonely or socially isolated. That's well over fifty-five million people—far more than the number of adult cigarette smokers and nearly double the number of people who have diabetes. A 2018 AARP study using the rigorously validated UCLA loneliness scale found that one in three American adults over the age of forty-five are lonely. And in a 2018 national survey by the US health insurer Cigna, one-fifth of respondents said they rarely or never feel close to people.

Consequences of Loneliness

There is growing evidence that loneliness has serious implications for our health. Dr. Julianne Holt-Lunstad of Brigham Young

University and her colleagues have found that lacking social connection is associated with a reduction in life span equal to the risk of smoking fifteen cigarettes a day—greater than the risk associated with obesity, excess alcohol consumption, and lack of exercise. A growing number of research reports have also found that loneliness is associated with a greater risk of coronary heart disease, high blood pressure, stroke, dementia, depression, and anxiety. Studies also suggest that lonely people are more likely to have lower-quality sleep, more immune system dysfunction, more impulsive behavior, and impaired judgment.

In addition to our health, social disconnection adversely impacts how we perform in school and in the workplace. It is also an important root cause of the polarized climate of distrust and division that hangs over much of the world. In the same way that healthy connections help us work through challenges in a relationship, strong human connections can help us work through societal challenges. Communities around the world are dealing with climate change, terrorism, poverty, and racial and economic inequities. Addressing these issues requires dialogue and cooperation. But even as we live with increasing diversity, it's easier than ever to restrict our contact, both online and off, to people who resemble us in appearance, views, and interests. And it's easy to dismiss people for their beliefs or affiliations when we don't know them as human beings. The result is a spiral of disconnection that's contributing to the unraveling of civil society.

It's a vicious cycle. When we're disconnected, we have a hard time listening to one another. We tend to judge quickly and assume the worst about people who disagree with us. This makes working together to overcome challenges increasingly difficult. The more problems we face, the angrier we get, which fuels the cycle of fear and distrust that stokes alienation and a sense of estrangement from society.

What is making so many people feel so disconnected? In our increasingly complex world, the causes are equally complex. While technology promises to connect us, it can also isolate. While mobility means our loved ones are only a train ride or flight away, we also move away from the communities where we grew up. While the pursuit of individual happiness is a right, we can put our own goals ahead of our relationships and community. Despite all of the progress we have made in how we talk about mental health, we are still held back by shame about our loneliness.

Causes of Disconnection

Of all the factors contributing to loneliness, the one that always seems to bubble to the top of mind for most people is technology. It is increasingly clear that it holds mixed blessings for us. Social media can help people find meaningful connections. But in the wrong circumstances, it can exacerbate loneliness by amplifying comparison, enabling bullying, and substituting lower- for higher-quality relationships.

There has been much public discussion about the amount of time people, especially children, are spending in front of screens. UK researchers Andrew Przybylski and Netta Weinstein found that the well-being of adolescents is optimized when their screen time is limited to one to two hours on a given weekday (interestingly, those with no screen time felt worse than those with modest use). Recent estimates, however, put the average time teenagers spend on screen-based entertainment at above six and a half hours a day, including watching videos and participating in social media.

Equally important is *how* we are using technology. A few minutes of harmful online content, for a susceptible person in the wrong circumstances, could be devastating, while an hour of screen time, as part of a rich experience with family and friends, might be very positive. As technology researcher Dr. Amy Orben puts it, "The problem is we're focusing so much on time spent on screens but not focusing enough on the content, type of technology, or motivation to use it."

To be sure, modern communication technology can bring us closer together in many ways. Social media can allow people who are isolated—due to disability or illness, or because they belong to marginalized groups—to find enriching communities. Such platforms can make it easier to reconnect and stay in touch with old friends. They also give us a way to more easily share important moments like the birth of a child or the loss of a loved one with our networks of friends, linking us to sources of support.

I remember as a child mailing blue single-paged aerograms to faraway relatives in India and filling every possible space on the paper. It took two weeks for the aerograms to be delivered, and we had to wait an additional two weeks or more for a reply to travel back halfway across the world. Today, thanks to video conferencing technology, my children can share a virtual meal anytime with their grandparents across the country. When I'm away on overnight trips, I can still admire my son's latest artistic creation or cheer on my daughter as she toddles around the house.

Dr. John Cacioppo, the founder of the field of social neuroscience and the expert most responsible for putting loneliness on the map as an important scientific concern, would say that the connecting power of technology is at its best when it serves as an online way station to connect people offline. While turning to one's social media feed as the destination can often leave people feeling more distant and dissatisfied with their own lives, the prosocial use of media platforms as a link to human engagement offline has been shown to decrease loneliness.

Stealing Time

As powerful as these benefits may be, the way that many technology platforms are designed—particularly social media—leads to use that ultimately detracts from our connection with one another. How often have you meant to spend five minutes checking your friends' posts only to end up spending an hour? You message a friend on Facebook, then bounce from profile to profile, checking out the cats, meals, and travels of people you barely know. We may tell ourselves these online forays are just diversions, but they're stealing time that we could be spending in real life with family and friends.

This theft of time is abetted by the seductive and dangerous myth of multitasking. Electronic devices—smartphones in particular—have promoted this myth like never before. All of a sudden, we can talk on the phone, send emails, pay our bills, order our groceries, and travel across town, all at the same time. It seems easy and efficient. It creates the illusion that we can satisfy our curiosity in a dozen directions at once—simultaneously hearing a friend's story about his new baby, checking out a neighbor's vacation photos, picking up a text about a parent's trip to the doctor, and googling the latest news about our favorite sports team. But when we multitask, we're splitting our attention into smaller and smaller fragments, reducing efficiency and diminishing the quality of engagement we bring.

Research has found that humans are incapable of attending to multiple activities at once. What we're actually doing when "multitasking" is switching back and forth very rapidly between tasks, attending to each one separately but briefly. As MIT neuroscientist Dr. Earl Miller explained in a 2008 interview on NPR, "Switching from task to task, you think you're actually paying attention to everything around you at the same time. But you're actually not."

In the middle of a conversation, for instance, when stealing a look at your cell phone, you may hear and remember the words that are spoken, but you won't process the words and nonverbal cues nearly as quickly or completely. One reason for this is that tasks involving communication compete for the same pathways within the brain. "Those things are nearly impossible to do at the same time," Miller said. "You cannot focus on one while doing the other." No wonder that the constant presence of our phones and other communication technology has been shown to reduce the emotional quality of our conversations. As Przybylski and Weinstein found in their experiments, the mere sight of phones during conversation negatively impacted "the extent to which individuals felt empathy and understanding from their partners."

Culture of Comparison

Distraction is not the only reason that technology can interfere with high-quality connections. As I found before paring back my Facebook usage, social media also fosters a culture of comparison where we are constantly measuring ourselves against other users' bodies, wardrobes, cooking, houses, vacations, children, pets, hobbies, and thoughts about the world. It's a bit like a continuous high school reunion, where everyone is "sharing" their accomplishments, victories, and delights, vying to prove their worth. Some may simply want to share joy with friends, but the net result often is a curated portrait of seemingly perfect lives, which in turn can make us feel anxious, depressed, and worse about ourselves by comparison. The most susceptible are young people, who are still in the process of defining their identities and goals.

When we're making comparisons online, we're not just rating ourselves. We're also comparing our various options in possessions, jobs, activities—and potential friends and partners. The digital pipeline presents us with a seemingly endless supply. Swipe left. Swipe right. Over and over and over. Certainty about our choices can quickly falter when the virtual supply chain promises to present us with ever better, brighter alternatives the next time we log on. Once we've selected our roommates, friends, and intimate partners, we'll have to do the messy offline work of getting to know one another's true complexity, and we might not love what we find. The allure of the "perfect" match, then, is a powerful deterrent to commitment. But perfection is an illusion that technology and modern culture cultivate at the expense of humanity. The perpetual cruising, the endless chase for the ideal companion is bound to leave us anxious and lonely.

Reduced Solitude

One of the less obvious sacrifices we make at the altar of technology is our time for solitude. Unlike loneliness, solitude is a state of peaceful aloneness or voluntary isolation. It is an opportunity for self-reflection and a chance to connect with ourselves without distraction or disturbance. It enhances our personal growth, creativity, and emotional well-being, allowing us to reflect, restore, and replenish. As technology has rapidly filled in the white space in our lives, our capacity for solitude has diminished. Social media's constant presence creates the illusion that we never need to be alone—and that something must be wrong with us if we *feel* alone. Yet we still need solitude, as well as the time and space to cultivate its benefits. We need regularly to free our minds to wander and explore without being directed by network algorithms and targeted ads. Solitude allows us to get comfortable being with ourselves, which makes it easier to *be* ourselves in interactions with others. That authenticity helps build strong connections.

To be real is to be vulnerable, and this can be challenging, especially if we believe that others will like us more if we hide or distort who we truly are. Technology can promote this belief by making it easy to pose online as someone braver, happier, better looking, and more successful than we really feel. These poses, in fact, are a form of social withdrawal. Although they may let us pretend that we're more accepted by others, our awareness of pretense only heightens our loneliness.

Addictiveness of Tech

In the end, technology is a tool that can be used to strengthen or weaken the social connections in our lives. What ultimately determines whether we reap its benefits or pay its price? One might think, isn't it the responsibility of the user to exercise judgment and willpower in moderating their use of technology? In theory, yes. But in practice, today's technology platforms are developed with a highly sophisticated understanding of human behavior and brain science. Software engineers use all manner of techniques—from autoplay on YouTube to streaks on Snapchat to interaction notifications on Instagram, Twitter, and Facebook—to keep bringing us back and to hold our attention on their platforms for as long as possible. In most cases, the economic measure of a successful app is not the quality of human interaction online but sheer quantity of usage. The more time we spend on the platform, the more revenue it generates, usually in the form of advertisements. In other words, our time is social media's money. In this way, apps have become the quintessential products of the attention economy. Drawing boundaries around our use of these platforms is not a simple matter.

Moreover, social media has woven its way into our social and professional lives. If you're a reporter, you can't afford to turn off Twitter entirely. If you're looking for a new job, having a profile and being on LinkedIn may be essential. If your family and friends use social media to announce major life events or get-togethers, and you're not on that platform, you can find yourself in the dark.

Creating a connected life begins with the decisions we make in our day-to-day lives. Do we choose to make quality time for people? Do we show up as our true selves? Do we seek out others with kindness, recognizing the power of service to bring us together? The answers to these questions have far-reaching implications once we understand that our connection to others has a fundamental impact on our health, our performance in the workplace, and our ability to overcome growing polarization in favor of constructive dialogue and mutual support.

The key question with technology is whether it influences these decisions in favor of connection or isolation. While there is little question that our lives have been transformed by technology, what is far less clear is the depth and breadth of that impact. When it comes to manufacturing drugs and medical devices, we rigorously assess harm and benefit. Yet despite the fact that technology from cell phones to email to social media has profound effects on our lives, we understand far less about the size and scope of impact.

For too long, technology companies have avoided grappling with the adverse social and health consequences of their products. This can no longer be the case. It is incumbent on technology companies and a new generation of humanistic entrepreneurs to imagine and design technology that intentionally strengthens our connections with one another instead of weakening them, that prioritizes quality in our interactions over quantity, and that supports a healthy, informed, and engaged society. And it is the responsibility of the public and policymakers to demand accountability from companies and support them when they act responsibly. The more our lifestyle evolves to maximize efficiency at the expense of human interaction, the more focused we must become in directing our use of technology to facilitate deeper personal connection.

Our connections with people are both our greatest source of fulfillment and also the ultimate performance enhancer. The way in which we are using technology may be detracting from our connections, but we have the ability to change that. We can be intentional about drawing boundaries around our use of technology while protecting sacred spaces for undistracted conversation. We can be mindful of when we are expressing less empathy and compassion in our online interactions and choose those moments to pause, pick up the phone, or visit someone to talk in real time. We can model people-centered living for our children, prioritizing in-person interaction, picking up the phone when loved ones call, and choosing to interact with colleagues, classmates, neighbors, and strangers as we move through the world instead of allowing our devices to consume all our attention. We can demand technology that supports human connection instead of platforms and services that seek to maximize and monetize our attention at all costs.

The work of building a more connected life is not easy. It requires courage—to be vulnerable, to take a chance on others, to believe in ourselves. But as we build connected lives, we make it possible to build a connected world. In such a world, we design our schools, workplaces, and technology to support human connection. We shape our laws to be forces for strengthening community. We treat kindness and compassion as sacred values that are reflected in our culture and politics.

The great challenge facing us today is how to build a people-centered life and a people-centered world. So many of the front-page issues we face are made worse by—and in some cases originate from—disconnection. Many of these challenges are the manifestation of a deeper individual and collective loneliness that has brewed for too long in too many. In the face of such pain, few healing forces are as powerful as genuine, loving relationships.

Adapted from Together: The Healing Power of Human Connection in a Sometimes Lonely World *(New York: Harper Wave, 2020).*

PART 2

HOW TECH IS HURTING KIDS

Kids Interrupted: How Social Media Derails Adolescent Development

MADELINE LEVINE

Madeline Levine, PhD, is a psychologist, clinician, consultant, educator, and author. She is the cofounder of Challenge Success, a program of the Stanford University Graduate School of Education, and the author of the New York Times *bestselling books* The Price of Privilege, Teach Your Children Well, *and* Ready or Not.

While the debate (and the research) continue about whether or not social media contributes to diagnosable mental illnesses like anxiety disorders and depression, a clearer, also troubling finding is more conclusive yet gets far less attention. Social media has been shown to make a contribution—a large one—to unhappiness. Given that we are in a period of rising rates of emotional problems for our kids (and all of us, actually), the fact that social media has a negative impact on mood is a not-to-be-overlooked, important finding.

Based on ongoing large-scale research, as well as my thirty-five years of clinical work, I would add that social media also has a negative impact on certain normative and necessary aspects of adolescent and young-adult development, particularly risk-taking, independence, and self-esteem. For many kids it also heightens a sense of isolation. Combine these findings, and we end up with large numbers of teens who are notably unhappy and thwarted in their development. Whether or not these facts lead to higher rates of mental illness remains to be seen. Robust research, which currently tilts toward confirming that social media frequently results in increased anxiety and depression in youth, will eventually clarify and quantify this question. In the meantime, there is enough damning evidence of social media's toxic effects on multiple factors that we need to step back and be more thoughtful about how our children use social media and what our responsibility is, and should be, around monitoring this activity.

It is important to bear in mind that for some kids, social media is a supportive and even life-saving experience. Kids who are geographically isolated, rejected by their community, or suffering from social phobias can benefit from the connections that social media encourages. Overall, however, social media can result in incessant negative self-evaluations because of compulsive comparisons with carefully cultivated but inauthentic images, as well as disruption in sleep, concentration, and physical activity. Given the runaway dependence of most teenagers on social media sites, we are at a point of much-needed reevaluation around the role and impact of social media on our kids' lives as well as our own.

Watching Out for the False Self

When kids feel perpetually pressured to perform brilliantly in the public realm and are rewarded for doing so, they have difficulty turning inward and being reflective. Overly dependent on the approval of others, including their peer group, teachers, and coaches, they have been so successfully trained to seek external affirmation that they have a hard time expressing or even recognizing what interests they actually enjoy, which friends they prefer, what really matters to them, where they stand. They're stuck in what psychologists call a "false self": Ivy League–bound! Possible baseball scholarship! Queen Bee!

Constructing a vibrant, but inauthentic, social media persona comes rather easily. Fifteen years ago, when I wrote *The Price of Privilege*, I opened the book with the story of my young high school patient who looked polished but was in fact despairing and had cut the word *empty* into her arm. The phenomenon of the false self is not new, and unfortunately, many of our kids have had way too much practice in creating one. A foundational task of adolescence is the development of a deep and authentic sense of self. My opinion is that our escalating rates of depression are, at least in part, due to forgoing this process and instead crafting a superficial sense of self built solely on the approval of others.

Who is this social media person? He or she is a stellar student/athlete/artist/musician/budding entrepreneur/social butterfly and also a compulsive chronicler of his or her triumphs. The correlation (correlation is *not* necessarily causation) between social media and adolescent depression and anxiety is well documented: The decrease in life satisfaction, self-esteem, and happiness among teenagers over the past decade correlates with the arrival of iPhones (in 2007), Instagram (2010), and Snapchat (2011) and with texting as the most prevalent form of communication (2007). As compared with adolescent boys, adolescent girls use mobile phones with texting applications more frequently and intensively. They are more prone to the compare-and-despair factor. Social media appears to have normalized a version of life that is actually achieved only by teams of makeup artists, hairdressers, and stylists. Instagram, the most image-driven social media platform, appears to lead the pack in ill effects on body image as well as heightened feelings of anxiety and depression.

The ill effects of social media are not limited to overt events such as online bullying, body shaming, or social exclusion. One colleague told me that his client, a fifteen-year-old boy, had vowed not to get involved with any girls in high school because of what he called "the audience." Girls would want to post selfies with him, he worried, which would then be rated in likes. The girls would be texting their friends, and the friends would weigh in on the boy's attractiveness, coolness, and whether or not he was texting the girl enough or saying the right things. Dating required a twenty-four-hour performance, no goofy comments or missteps allowed. As we know, the adolescent brain is painfully attuned to social rejection.

The public exposure completely unnerved this young man and was delaying his social development in the arena of boy-girl relationships. I have to admire his courage, though: He managed to figure out his own values despite intense peer pressure. What about the girls? They're subject to exactly the same audience, with even greater scrutiny applied to their appearance. As they become more engaged with their online selves and less involved with real-life relationships, their development gets impaired, too. A preoccupation with one's social media "self" injures self-esteem, cultivates FOMO (fear of missing out), and delays and distorts the development of an authentic self.

Increase in Social Isolation

Whenever I've spoken at schools, I always ask middle- and high-school students what troubles them most about being a teenager. I have them rank three common adolescent issues: developing an identity, feeling that they can make choices for themselves, and feeling isolated. Initially surprising to me, but by now expected, the vast majority of kids list social isolation as their biggest concern. For years before that, identity—the previously defining issue of adolescence—came in first.

Technology is at the root of much of this isolation, and the reasons are complicated. There are aspects of tech, such as gaming, that provide lively community, and the internet offers plenty of options for learning, creativity, and political action. As seen above, the problem is "the audience." For teenagers, especially twelve- to fifteen-year-olds, the only thing that matters is what their friends think. All the traditional horrors of being fourteen are still true: Do you sit with the popular kids? Is your girlfriend going to steal your boyfriend? Do you have a zit? You call those boobs? But instead of being segmented into specific and predictable parts of the day—lunch, the walk home from school, a phone call or two, a whispered taunt when the teacher's back is turned—opportunities for humiliation are now constant, thanks to texts and social media.

There's also a subtler element driving the isolation. Texting makes it easy to be hard—easy to break up with someone without looking them in the eye and seeing the hurt on their face. It's not uncommon in my practice for kids to walk into my office crying and hand me their phone. The messages I've read in these situations vary from the merely inconsiderate to the outright treacherous. That's bad enough, but commiseration has also been relegated to text. When your girlfriend breaks up with you, six texts from your buddies calling her a bitch doesn't help all that much. You need a friend beside you; you need the familiar sight of them and the feel of their shoulder bumping against yours as they help you walk off your blues. Historically, friends could spend hours unraveling hurt, providing support, and simply being a reminder by their presence that all is not lost. Today's kids are busy and have become accustomed to superficial contact. Instead of real consolation, all too often the most a friend can offer is a text full of emojis.

Face-to-face encounters are where children and adolescents learn empathy, diplomacy, and how to listen. Without tough or even heartbreaking real-life conversations, they get less practice at human interaction. In the coming years, which will bring frequent job changes and more collaborative work environments, people who have superior interpersonal skills will be at an even greater advantage than they are today. Those who lack such skills will have a harder time not only forming connections at work but also sustaining relationships with friends and romantic partners.

Disrupted Sleep, Concentration, and Competence

Often parents would come see me about the possibility of getting accommodations at school for their children. Sometimes this was done in the hopes of giving their children a "leg up" via extra time on standardized testing, but they often requested it because their kids were spending five and six hours a night on homework. They were concerned that perhaps a learning disability of some sort was impacting their child's ability to complete work in a reasonable amount of time. Typically, after observing their teen and having them keep a media diary, it became clear that the only thing slowing their work was their toggling back and forth between homework and social media. For example, one child who consistently and adamantly reported using ten to fifteen minutes of social media during homework was genuinely surprised that, in fact, he was spending over an hour online. While many of us are convinced we are extraordinarily good at multitasking, the research is clear we aren't. Switching back and forth between tasks is inefficient, tiring, and impairs our concentration, especially when one task can have a large emotional component. Three hours of homework becomes five, and the work is unlikely to be your child's best.

Gaining competence is one of the main tasks of life. Competence is the result of interest, talent, and practice. Teenagers in particular should be busy practicing a range of activities. At the simplest level, they have to learn how to walk in their newly acquired bodies without bouncing off walls. They also have to learn how to talk to the opposite sex, feel relatively at home in their ever-changing bodies, find things that interest them, and take a deep dive into their goals, values, and hoped-for future. This process takes time and energy. The young woman with "empty" on her arm personified how excessive attention to external validation interferes with the critical tasks of identity formation and competence. This was true long before social media, which has intensified the preoccupation with the judgment of others.

When kids spend three or four hours a day on social media, what other parts of life are neglected? Physical activity? Family time? Face-to-face time? Restorative time? Practice time? Contemplation about one's talents, interests, hopes, and dreams? Completing adolescence has always been both an exhilarating and a daunting task. It is unlikely to be completed with the same effectiveness with fewer hours in the day.

Finally there is the issue of how screen time in general, and time on social media in particular, interferes with sleep. Kids I've seen in the last few years are almost invariably sleep-deprived. They push themselves, and their parents push them, hard. They're encouraged to take multiple AP classes. Stellar academic performance is expected. But there simply are not enough hours in the day. Often I'll do a pie chart of a teen's day with both parents and the teenager. They are often shocked to see that the teen is stuffing thirty hours of work into a twenty-four-hour day. That results in lots of multitasking and little sleep. Here's the reality: Teenagers absolutely need a good night's sleep—a minimum of eight hours, and preferably nine. We know that sleep deprivation is a contributor to depression, and too many kids have gotten into the habit of using drugs like Adderall to stay awake and complete their work. Whether or not naval aviators sometimes use Adderall to stay awake during extended missions is irrelevant (I've heard this rationalization too many times to count). Our kids are not fighter pilots, and Adderall is not risk-free. Relying on pills to enhance comfort and capacity is not a road you want your child to go down.

Most kids are using their phones and on social media before they go to sleep. Research shows that this has multiple deleterious effects. The blue light from electronic screens makes it harder to fall asleep, and kids who keep their phones with them in bed—checking how many "likes" they have, how they compare to friends, or what the latest bit of social drama is, which are all conducive to stress—get an hour less sleep a night than kids who don't take their phones to bed. Instead of relying on phones to wake them up, kids should use alarm clocks. Don't let your kid take his or her phone to bed.

What to Do?

Given the documented increase in unhappiness and loneliness—and the likelihood of increasing depression and anxiety—associated with the use of social media among teens, it is critical that parents put limits on usage. It may feel good to have our kids at home, know where they are, and what they're doing. But lying in bed, alone for hours at a time with only their phones for companionship, comparing themselves to enhanced versions of other kids, they have less time to work through the developmental tasks of risk-taking, independence, identity, intimacy with others, and moral development. Jean Twenge's research on almost half a million middle and high schoolers suggests that anything over two hours a day of time on YouTube, Instagram, Snapchat, Facebook, or whatever platform has since grabbed interest has detrimental effects on kids' mental health. We also know that over half of teens describe themselves as being on social media or the internet "almost constantly."

To counteract the negative effects of "almost constant" screen time, parents can do several things:

- Make sure there is balance in your kid's life.
- Model it in your own.

- Allow for some social interaction online, but the majority should be offline. To learn the fine points of social relationships, kids need all the cues they can get. That includes subtle facial and physical cues you can't get on a smartphone.
- Make sure your kid's school does some digital media literacy training. If it doesn't, talk to the administration about adding this.
- Cut down device use by recruiting kids to do their own research on digital and social media. Kids are curious and hate to be manipulated. Discovering the manipulation behind their attachment to devices is a strong deterrent. Look at how drastically cigarette smoking rates among teens dropped when the manipulation of cigarette manufacturers was exposed.
- Shut off notifications, and have your kid do the same. Notifications are just one of the many ways that you're manipulated into constant interaction with your phone.

For kids who have been online from birth, setting limits on their usage is bound to be profoundly distressing. Parents, however, are not in a popularity contest. In the same way we do our best to ensure our kids' physical health with exercise, sleep, and nutritious food, we also have to guard their mental health, even if it means tolerating some eye-rolling and slammed doors. If you can't stand to see your kid unhappy, you're in the wrong business.

Put down your phone for most of the day, and insist that your kids do the same. Having lived through COVID-19, a time of unprecedented fear and isolation, it should be clear that there is much to be gained from our phones, but much more to be gained by being present, involved, and in touch with the people around us.

Big Tech Failed My Generation

CAMERON KASKY

Cameron Kasky is an American activist and advocate against gun violence. After surviving the mass shooting at Marjory Stoneman Douglas High School in Parkland, Florida, in February 2018, he cofounded the student-led gun violence prevention advocacy group Never Again MSD and helped organize the nationwide student protest March for Our Lives in March 2018.

Social media can be a tool for good. When a massacre happened at my high school in early 2018—taking seventeen lives, injuring seventeen more, and causing lifetimes of trauma and pain for thousands of others—social media is where my classmates and I went to call for change. We used it as a tool to communicate our message to people across the country and the world. We weren't going to sit back and accept the typical mass-shooting narrative. We weren't going to let the cameras come to town, film a bunch of crying mothers, and talk about how the shooter was just a misunderstood, well-meaning boy who deserved better. We told the country that it is all of our responsibilities as American citizens to defend our country from violence in our schools, movie theaters, places of worship, and everywhere else, and the only way to do that is to look at the constant in these tragedies: the weapon of war that is used to carry out an overwhelming number of them.

Without social media, it would have been practically impossible to connect dedicated change-makers around the country and organize hundreds of marches around the world, where millions of people demanded fundamental change. On top of that, social media has been used as a platform for many young candidates in elections around the country who do not have the overwhelming amounts of money that their opponents might to make their voices heard. On the mental health side, programs like the Trevor Project use social media to help save the lives of LGBTQIA+ people and to make an inspiring difference through their services and advocacy. Social media will never go away, and many more great things have come out of it.

But let's be real. As a nineteen-year-old, I am part of a generation shaped by relationships with technology that are not too far from what was once considered dystopia. Witnessing tech's oppressive grasp on our psyches, I am thoroughly convinced that we are plummeting to depths from which we will never recover. Without my smartphone on me at almost all times, I feel as though part of my body is missing, and I simply cannot move on without it. According to my beloved iPhone—the controller in this toxic relationship I've always had with my devices—I put in around six hours of screen time a day. My phone is almost essential for the work I do both academically and professionally—but at least four of those hours are spent scrolling through social media to look at superficial versions of my friends' lives. I use my iPhone until my eyes hurt and I have to lower the brightness. I sit and refresh social media apps for new content mere seconds after my last venture into the abyss. I frequently fall asleep with it in my bed. And I might add, I use my phone significantly less than many of my peers.

I would be lying if I told you that I know much about how smartphone companies manipulate our brains with their devices, but I am sure that they make plenty of efforts to get us to use their products as much as possible, with no regard for the effects this might have on us. My generation thinks little about this and enjoys the blissful ignorance of consuming media like candy. Candy is, in all honesty, probably a bit healthier than our relationships with mobile devices. We certainly do raise eyebrows when our voice-control programs pick up things we are saying when we haven't even turned them on. But who cares? My Siri has an Australian accent and I think it's funny. So I let it be. I don't know how it works. I simply know *that* it works and works exceedingly well. I can't put the cursed thing down even when I try. I try very often.

From my point of view and essentially zero research, the real terror is social media. There is a massive and powerful threat in cyberbullying. It is the direct cause of countless suicides. It plants the seeds of fundamental self-doubt earlier and earlier, as social media users continually get younger. I deleted Facebook after receiving thousands of violent death threats (although I should've deleted it before Zuckerberg got all my private information to play with). But the cyberbullying I constantly face does not affect me the way that it affects many others. I find right-wing trolls attacking me over my support for gun-control policies objectively hilarious. But that's just me. The effects of cyberbullying are traumatic and destructive to many, many people of all ages.

I believe it is also important to talk about the casual, laid-back psychological tragedies on the apps we use to connect. Friends of mine base their self-worth completely on the comments they receive on the pictures they post and on the followers they accumulate. Strong, inspiring people I love and respect see replies to their posts that make them cry. Social media is detrimental to the way we see ourselves even when there aren't anonymous users in our inboxes telling us to kill ourselves. This is—to simplify a multifaceted and unresolved issue—because social media and real life have been conflated in the exact ways that many have feared for a long time.

Social media is an opportunity to project a version of yourself that you want the world to see and make this version of yourself accessible to many, many others. As exciting as this may sound, millions of teenagers and adults alike begin to see others' social media presences as their authentic selves. If Bill sees his friend Jim post a picture in which his face is edited to look flawless and he is surrounded by loving and excited friends, Bill will likely convince himself that this is the life Jim lives. Bill will inevitably compare himself to his friend Jim. "Why can't I be like that? Why can't I live that life and be happy like Jim?" This crisis is not the result of talking points perpetuated by baby boomers and cynics. This is a real-life crisis that far too many of us convince ourselves is not our problem.

The self-worth and confidence of young people everywhere is being undermined. Meanwhile, our information is getting stolen, shuffled around, and sold like trading cards among powerful marketing entities. The neurological, psychological, and emotional development of too many of us who can't put these stupid devices away is being threatened.

To the big tech companies that pretend they are taking action to foster safer relationships between my generation—and every generation—and our technology, you are continuing to fail us. You have exploited us and caused damage, the likes of which are yet to be fully determined. Until actual changes are made, you are on the wrong side of history, and history will not forgive you.

The New Normal

WILLOW BAY

Veteran broadcast journalist and new media strategist Willow Bay is dean of the Annenberg School for Communication and Journalism at the University of Southern California.

For Pedro Resendez—a sixteen-year-old in Mexico City—incorporating a mobile device into his daily life as an early teen was a challenge and a source of tension between him and his parents. But over time, that family dynamic shifted. "I think my parents started trusting me more," he said. "They know that I use the phone as a positive rather than a negative tool."

Like other teenagers, Resendez is on the leading edge of a massive, global transformation, unprecedented in its speed, scope, and breadth. The diffusion of wireless technologies, the accompanying advances in digital media and mobile devices, and the rising power of social media are changing the way we engage not only with the world around us but with the people who are closest to us. Patterns of daily life have been forever altered by the ubiquity of digital devices—and no one left us a user's manual.

These shifts are having an impact on people of all ages. But digital natives, also known as Gen Z (born approximately between 1997 and 2012), are on the leading edge of this profound transformation—experiencing life through these communication technologies, not merely on them.

Teens today are connected 24/7. Their emotional and social development, as well as their academic and intellectual development, is inextricably linked to the rapid advances and deep penetration of technology in their lives. As seventeen-year-old British student Sophia Hodgson noted, "We live in a modern world of technology where everything is instant," but she'd feel "relieved" without the "constant pressure" of needing to respond to texts, emails, and other notifications.

This generation is emerging as the most plugged in, connected, and socially conscious generation the world has ever seen. Their "always on" culture brings with it really extraordinary advantages, but it's also left many of them sleep-deprived, distracted, and overly dependent on their devices. "Some of us are unconscious of being on the phone a lot," Japanese high school student Riyo Kombashi explained. "Because it's such a natural daily habit, you're not even aware of it."

As we experience perhaps the first technological revolution with teens leading the way, their parents—and I am one of them—are the first generation to have to manage mobile devices in their own lives, in their own hands, and in the lives of their children.

In 2017, USC's Annenberg School for Communication and Journalism joined forces with Common Sense Media to develop a global research framework and generate meaningful data on the impact of mobile devices on our family relationships.

Our effort, *The New Normal: Parents, Teens, and Mobile Devices around the World*, is a multiyear research collaboration designed to advance cross-cultural exploration of family digital media engagement—specifically, the ways in which parents and teens are adopting and adapting to mobile devices in their lives and how they view each other's device use. The goal is to encourage thoughtful conversations to help us understand both the benefits of this transformative technology and the downsides in the lives of our children and families.

In 2016, Common Sense published its first US reports on this topic, titled *Technology Addiction: Concern, Controversy, and Finding Balance* and *The Common Sense Census: Plugged-in Parents of Tweens and Teens.* They revealed a new family dynamic driven by tech and shaped by its benefits and drawbacks. These reports became our benchmark for comparison across countries when we released a series of joint studies: *The New Normal: Parents, Teens, and Mobile Devices in Japan* (2017), *The New Normal: Parents, Teens, and Mobile Devices in the United Kingdom* (2018), and *The New Normal: Parents, Teens, and Mobile Devices in Mexico* (2019).

Each report is based on representative online surveys of more than twelve hundred participants, consisting of parent-teen pairs. The surveys address mobile device use, habits, and attitudes with questions concerning distraction, feelings of "addiction," and family conflict, as well as the benefits of mobile technology. We also host conversations with local teens—invariably the most compelling part of our work. They share their thoughts and feelings about the role of tech in their family life, and they often surprise us with their insight, awareness, and nuance.

From the research, it is clear that mobile devices are reshaping daily life for parents and teens in all the countries we have studied:

- Parents and teens are on their mobile phones early and often.
- The vast majority of parents and teens are distracted daily by their mobile devices.

- In many families, mobile devices are interrupting sleep.
- Most parents and teens feel that teens' use of mobile devices interferes with family activities.
- Nearly half of parents and teens report feeling addicted to their mobile devices.
- Most parents feel that their teen spends too much time on their devices and express concern about their own feelings of addiction.

Most families also agree on the benefits, such as acquiring important tech skills, staying in touch with extended family, and keeping up with current events. Their feelings of optimism about the benefits of technology sit alongside feelings of distraction and addiction. Across all four countries, however, the majority of parents and the vast majority of teens say that mobile devices have made no difference to their family relationships. In fact, for some, adopting and adapting to this changing world has become a shared experience. "Parents are new to this technology just like we are," observed Ava Hall, a senior at the American School in Japan. "Right now, we are both technically 'growing up' with technology, both parents and kids."

Listening to teens around the world examine their own habits and reflect upon what they observe about their parents is a critical step toward a deeper understanding of the impacts of mobile device use. In these conversations, we have observed teens express a growing awareness about their relationship with their mobile devices, perhaps leading to a more mindful integration of technology into their lives.

At this extraordinary moment in history, we hope to offer timely, relevant data that will encourage further interest and research, locally and globally. The greatest value of this work, however, may be the conversations it inspires with policymakers, technologists, media professionals, educators, and most of all, families themselves.

Why Section 230 Hurts Kids and What to Do about It

BRUCE REED AND JAMES P. STEYER

Bruce Reed served twelve years in the White House as President Bill Clinton's chief domestic policy adviser and Vice President Joe Biden's chief of staff. As senior adviser at Common Sense Media and CEO of Civic, a bipartisan policy ideas company, he helped write the landmark California Consumer Privacy Act of 2018. James P. Steyer is founder and CEO of Common Sense Media.

Mark Zuckerberg makes no apology for being one of the least-responsible chief executives of our time. Yet at the risk of defending the indefensible, as Zuckerberg is wont to do, we must concede that, given the way federal courts have interpreted telecommunications law, some of Facebook's highest crimes are now considered legal. It may not have been against the law to livestream the massacre of fifty-one people at mosques in Christchurch, New Zealand, or the suicide of a twelve-year-old girl in the state of Georgia. Courts have cleared the company of any legal responsibility for violent attacks spawned by Facebook accounts tied to Hamas. It's not illegal for Facebook to allow posts to foment attacks on refugees in Europe or try to end democracy as we know it in America.

On the contrary, there's a federal law that actually protects social media companies from having to take responsibility for the horrors that they're hosting on their platforms. Since Section 230 of the 1996 Communications Decency Act was passed, it has been a get-out-of-jail-free card for companies like Facebook and executives like Zuckerberg. That twenty-six-word provision hurts our kids and is doing possibly irreparable damage to our democracy. Unless we change it, the internet will become an even more dangerous place for young people, while Facebook and other tech platforms will reap ever-greater profits from the blanket immunity that their industry enjoys.

It wasn't supposed to be this way. According to former California Representative Chris Cox, who wrote Section 230 with Oregon Senator Ron Wyden, "The original purpose of this law was to help clean up the internet, not to facilitate people doing bad things on the internet." In the 1990s, after a New York court ruled that the online service provider Prodigy could be held liable in the same way as a newspaper publisher because it had established standards for allowable content, Cox and Wyden wrote Section 230 to protect "Good Samaritan" companies like Prodigy that tried to do the right thing by removing content that violated their guidelines. But through subsequent court rulings, the provision has turned into a bulletproof shield for social media platforms that do little or nothing to enforce established standards. As Jeff Kosseff wrote in his book *The Twenty-Six Words That Created the Internet*, the provision "would come to mean that, with few exceptions, websites and internet service providers are not liable for the comments, pictures, and videos that their users and subscribers post, no matter how vile or damaging."

Facebook and other platforms have saved countless billions thanks to this free pass. But kids and society are paying the price. Silicon Valley has succeeded in turning the internet into an online Wild West—nasty, brutal, and lawless—where the innocent are most at risk. The smartphone and the internet are revolutionary inventions. But in the absence of rules and responsibilities, they threaten the greatest invention of the modern world: a protected childhood. Since the nineteenth century, economic and technological progress enabled societies to ban child labor and child trafficking, eliminate deadly and debilitating childhood diseases, guarantee universal education, and better safeguard young children from exposure to violence and other damaging behaviors. Technology has tremendous potential to continue that progress. But through shrewd use of the irresponsibility cloak of Section 230, some in Big Tech have turned the social media revolution into a decidedly mixed blessing.

Although the United States has protected kids by establishing strict rules and standards on everything from dirty air and unsafe foods to dangerous toys and violence on television, the internet has almost no rules at all, thanks to Section 230. Kids are exposed to all manner of unhealthy content online. Too often, they don't even have to seek it out; harm comes looking for them. Social media platforms run inappropriate ads alongside content that kids watch. Platforms popular with children are overrun with advertising-like programming, such as unboxing and surprise videos. Because their business model depends on commanding as much consumer attention as possible, companies push content to kids to keep them on their platforms as long as possible. All the tricks of manipulative design that make Big Tech dangerous for society—autoplay, badges, and likes—put young people at the greatest risk. In the early days of the web, a *New Yorker* cartoon showed a dog at a desktop, with the caption, "On the internet, nobody knows you're a dog." On today's internet, nobody cares if you're a kid.

Exhibit A: YouTube

Big Tech's browse-at-your-own-risk ethos is particularly evident on sites like YouTube, where kids are doing exactly what they've done for more than half a century—staring at a screen—with one key difference. There are no longer any limits on what they can watch. Google's algorithms profess to know everything we desire, but they certainly don't know what we want for our children. In fact, grown-ups are currently leading a wave of nostalgia for America's golden age of children's entertainment: *Sesame Street* celebrated its fiftieth anniversary; Tom Hanks starred as Mr. Rogers in a critically acclaimed movie; and the launch of Disney+ turned Disney's vast library of animated and adventure classics into the most successful streaming service of all time.

Such nostalgia is both understandable and ironic, when today's young kids are watching YouTube, an online channel that under federal law is not supposed to reach children under thirteen without parental consent. A Pew Research Center survey found that four out of five parents with children age eleven or younger let them watch YouTube, and a third of them watch regularly. Meanwhile, three out of five YouTube users say they come across "videos that show people engaging in dangerous or troubling behavior." Likewise, three out of five parents who let their young children watch YouTube say they encounter content "unsuitable for children." As the channel's proud parent, Google, has routinely boasted to advertisers, YouTube is the "new Saturday morning cartoons" and "today's leader in reaching children age six to eleven against top TV channels."

What might kids find on YouTube? YouTube videos aimed at kids have shown all manner of violence and perversion, from Peppa Pig armed with guns and knives to sex acts with Disney characters like Elsa. The Maryland couple behind FamilyOFive, a once-popular, now-terminated YouTube channel that attracted 175 million views, posted viral prank videos of child abuse against their own children. Perhaps most troubling, YouTube's behavioral algorithms appear to steer children into harm's way. An exhaustive research study funded by the European Union found hundreds of disturbing videos, with hundreds of thousands of views, offered up to children between the ages of one and five. The report concludes, "Young children are not only able, but likely to encounter disturbing videos when they randomly browse the platform starting from benign videos." Kids are growing up in the darkest age of children's entertainment in American history. As technology writer James Bridle warned in 2017, "Someone . . . is using YouTube to systematically frighten, traumatize, and abuse children, automatically and at scale, and it forces me to question my own beliefs about the internet, at every level."

The YouTube saga shows the folly of self-regulation when the laws aren't just weak but actually immunize companies from accountability for their behavior. Section 230 not only fails to protect kids from disturbing content, it also limits the effectiveness of other child-protective laws. In 2019, the Federal Trade Commission and the New York Attorney General went after Google for violating the Children's Online Privacy Protection Act (COPPA), which is supposed to prevent companies from collecting information from and personally targeting kids under thirteen. For all its limitations, COPPA was intended to give parents peace of mind and create a walled garden where children could not be preyed upon. Section 230 is a bulldozer that knocks those walls down, enabling platforms that profit off kids to avoid taking full responsibility for their actions. Many platforms skirt those provisions by claiming they do not have "actual knowledge" that users are under thirteen, as the law requires—even though they can usually gauge users' age from their online behavior. Google settled the COPPA complaint by agreeing to a modest $170 million fine.

What to Do about It

How can America revoke Big Tech's free pass before it's too late? First, we must set aside the industry's self-serving defense of Section 230. Platform companies insist that if they have to play by the same rules as publishers, individuals' right of free speech will vanish. But treating platforms as publishers doesn't undermine the First Amendment. On the contrary, publishers have flourished under the First Amendment. They have centuries of experience in moderating content, and the free press was doing just fine until Facebook came along. Section 230 is more like the self-protection that gun manufacturers—the only other industry in America with broad legal immunity—extorted from Congress under the pretense of the Second Amendment. The Protection of Lawful Commerce in Arms Act of 2005—passed just as the federal assault weapons ban expired—protects gunmakers from liability for crimes committed with their products. Hunters and gun owners don't benefit from that law, but it has unleashed the gun industry to sell millions of assault rifles with impunity.

The tech industry's right to do whatever it wants without consequence is its soft underbelly, not its secret sauce. Admitting mistakes is the sector's greatest failing, taking responsibility for those mistakes its gravest fear. Zuckerberg leads the way by steering into every skid. Instead of acknowledging Facebook's role in the 2016 election debacle, he slow-walked and covered it up. Instead of putting up real guardrails against hate speech, violence, and conspiracy videos, he has hired low-wage content moderators—by the thousands—as human crash dummies to monitor the flow. Without that all-purpose Section 230 shield, Facebook and other platforms would have to take responsibility for the havoc they unleash and learn to fix things, not just break them.

Congress never intended to give platforms a free pass. As Jeff Kosseff, the law's self-proclaimed biographer, points out, Congress enacted Section 230 because "it wanted the platforms to moderate content." So the simplest way to address unlimited liability is to start limiting it. In 2018, Congress took a small step in that direction by passing the Stop Enabling Sex Traffickers Act (SESTA) and the Allow States and Victims to Fight Online Sex Trafficking Act (FOSTA). Those laws amended Section 230 to take away safe harbor protection from providers that knowingly facilitated sex trafficking.

Congress could continue to chip away by denying platform immunity for other specific wrongs like revenge porn. Better yet, it could make platform responsibility a prerequisite for any limits on liability. Boston University law professor Danielle Citron and Brookings Institution scholar Benjamin Wittes have proposed conditioning immunity on whether a platform has taken reasonable efforts to moderate content. In their article "The Internet Will Not Break: Denying Bad Samaritans Section 230 Immunity," they note that "perfect immunity for platforms deliberately facilitating online abuse is not a win for free speech because harassers speak unhindered while the harassed withdraw from online interactions." Citron argues that courts should ask whether providers have "engaged in reasonable content moderation practices writ large with regard to unlawful uses that clearly create serious harm to others."

Demanding reasonable efforts to moderate content would represent progress. But that is a dangerously low bar for an industry whose excuse for every failure has been "sorry, we'll do better next time." A social media platform like Facebook isn't some Good Samaritan who stumbled onto a victim in distress. It created the scene that made the crime possible, developed the analytics to prevent or predict it, tracked both perpetrator and victim, and made a handsome profit by targeting ads to all concerned, including the hordes who came by just to see the spectacle.

Washington would be better off throwing out Section 230 and starting over. The Wild West wasn't tamed by hiring a sheriff and gathering a posse. The internet won't be, either. It will take a sweeping change in ethics and culture, enforced by providers and regulators. Instead of defaulting to shield those who most profit, the United States should shield those most vulnerable to harm, starting with kids. The "polluter pays" principle that we use to mitigate environmental damage can help achieve the same in the online environment. Simply put, platforms should be held accountable for any content that generates revenue. If they sell ads that run alongside harmful content, they should be considered complicit in the harm. Likewise, if their algorithms promote harmful content, they should be held accountable for helping redress the harm. In the long run, the only real way to moderate content is to moderate the business model.

In 2019, before Patrick Crusius massacred twenty-two people in an El Paso Walmart, he wrote a four-page white supremacist manifesto decrying a "Hispanic invasion of Texas." Like John Timothy Earnest—the disgruntled anti-Semite who opened fire on a synagogue in Poway, California, the same year—Crusius posted his racist thoughts on an online message board called 8chan. In March 2019, the shooter in Christchurch, New Zealand, streamed his killing spree for seventeen minutes on social media for millions to see. All three of those attacks (and others like them before and since) spread across the globe—inciting violence, glorifying white supremacy, and aggrandizing murderous young men intent on passing the torch of hate on to the next generation.

One crucial difference sets the Christchurch incident apart. In the wake of the El Paso and Poway shootings, Washington did what it has done so many times before—nothing. But New Zealand Prime Minister Jacinda Ardern won the world's heart by not only banning the military-style assault weapons the shooter used but by setting out to take away his other weapon: the spread of extremist content online. She challenged leaders of nations and corporations around the world to join the Christchurch call to action to make sweeping changes in laws and practice to prevent the posting and to hasten the removal of hateful, dangerous content on social media platforms. New Zealand could reform its gun laws, but, she said, "we can't fix the proliferation of violent crime online by ourselves."

In the end, Section 230 of the Communications Decency Act is no longer a necessary evil that nascent internet companies depend on to thrive. Instead, it has become our collective excuse to not take away the platform that hate depends on to grow and spread. The longer we do nothing, the more our humanity looks stripped, beaten, and half dead on the side of the road. Our kids know the moral to the story—Good Samaritans would stop to help the victim. So should we.

Raising Our Children in Two Worlds

SISSI CANCINO

Sissi Cancino is a Mexican journalist, communication consultant, and lecturer on issues of educational innovation. In 2017, she founded MODERS, an initiative that seeks to strengthen mothers as leaders of social change.

Being a mom is the most incredible thing that has ever happened to me. That said, I must add: It is complicated. It has always been.

Yes, love is the most important ingredient in the recipe of motherhood. But there are other essential elements: patience, instinct, principles, values, habits, time, disposition, dedication, passion, sleeplessness, and the need for continuous learning. Today, technology joins the list. Through its thousand forms—laptops, tablets, mobile phones, consoles, video games, smart screens, portable video games, social networks, and applications—technology has changed our family lives: how we communicate, what we learn and how we do it, how we socialize, and how we have fun.

For parents, technology should not and cannot be an enemy. If we play against it, it will undoubtedly beat us. Rather, by challenging ourselves to develop skills in this technological world, we can accompany and guide our children and boost their growth and development.

When I see my nine-year-old son, I know that he lives in a world so different from mine—one that is hyperstimulating, volatile, fast, and immediate. Apps allow us to access information and get quick results for everything: where to buy things, how to get somewhere, what to see when we travel, how to locate a place, where to eat, and how to pay. If I want to safely navigate this reality, I know that, as a mom, I need to develop and strengthen my own technological tools. I also know that I cannot give up spending time together, bonding, and teaching my son to become a gentle, kind, loving human being who is capable of becoming an agent of positive change for his environment, his ecosystem, and his time.

"Centennials" are pragmatic. They like to be self-taught (via YouTube, quite often) and trust collective knowledge construction. They are almost constantly in touch with the world through others but are increasingly disconnected from their personal and familial environments. They are prey to a culture of immediacy that makes it difficult for them to handle frustration and learn the value of patience and resilience. Children stuck to devices spend less time on physical activity.

How do we interact in a world where no one looks at one another and everyone is looking at their phones?

In 2019, a joint research initiative between Common Sense Media and USC's Annenberg School for Communication and Journalism published *The New Normal: Parents, Teens, and Mobile Devices in Mexico.* The study questioned more than a thousand parents and teens in Mexico about their mobile device use and digital media habits.

Sixty-seven percent of Mexican teenagers said they use their cell phones all the time. Fifty percent of teens feel addicted to their mobile phones. Mexican adults are not far behind: 45 percent of adults surveyed said they feel addicted to their phones.

We are not using our cell phones. We are living through them.

I want to live in a world that recognizes not only the reality and importance of technology but the need to continue raising our kids in love, independence, and trust, in a world where technology, mobile devices, and social media are just a part of life.

I want to live in a world that finds ways to adapt, without forgetting the truly important part of our role as parents: being there, connecting, looking our children in the eyes, talking about how they feel, and giving them the tools that will form them into healthy, independent, and responsible adults. With or without technology, we want them to find their voice and help give it strength, intention, and service. We want to respect their authenticity and essence and help them learn lessons from the real world that they can use, with the help of technology, to change both the real and the digital world for the better.

Using Technology to Boost Kids' Brain Development

CHELSEA CLINTON

As an advocate, author, and cochair of the Clinton Foundation's Too Small to Fail initiative, Chelsea Clinton is a champion for children and families. She is an adjunct assistant professor at Columbia's Mailman School of Public Health and the author of several books, including the New York Times *bestseller* She Persisted: 13 American Women Who Changed the World.

My first computer was a Christmas gift from Santa Claus in 1987, when I was seven years old. I used it mainly to play *Where in the World Is Carmen Sandiego?* and other games, and I thought it was the coolest thing in the world.

In the thirty-plus years since then, technology has infused almost every part of our lives. Today, 95 percent of American families with a child under nine in the home own a smartphone, and many children have access to screened devices, practically from infancy. Further, children as early as two are spending an average of three-quarters of an hour on screens, and that number jumps to more than two and a half hours a day among two- to four-year-olds and about three hours a day among five- to eight-year-olds, despite the fact that the American Academy of Pediatrics recommends no screen time for children under two, and no more than one hour a day of high-quality programming up to the age of five.

Much of today's technology—particularly when it's experienced through screens—is designed to distract and absorb attention in ways that we know are not conducive to healthy brain development. And there is much (well-deserved) hand-wringing about the correlation of too much screen time with kids' lower scores on cognitive and developmental assessments, higher obesity rates, disrupted sleep, and lower ability to read human emotions—not to mention concerns around exposure to violence online and the cyberbullying that pervades social media.

Yet technology is an integral part of most people's lives and plays a crucial role in how many families function—helping keep schedules organized and synced up, allowing parents and caregivers to keep an eye on sleeping infants, enabling them to communicate with older children at any time by texting, and providing a platform for kids to learn. So, we have to ask ourselves, how can we use technology in ways that benefit, and don't harm, children and their caregivers?

Research shows that technology can be a tool to support children's health and help them grow into healthy, happy adults. At the Clinton Foundation, we believe that every child deserves the best chance at success in school and in life, and that all caregivers have the power to lay the groundwork for that success. In 2013, the foundation launched Too Small to Fail, an initiative to give parents and caregivers information to empower them to boost young children's brain development in the first five years. Eighty percent of children's brains are formed by the time they're three years old, so it's hugely important to surround children with language from birth to provide them with robust cognitive scaffolding. Even simple, everyday interactions with young children—describing objects seen during a walk or bus ride, singing songs, and telling stories—can build their vocabularies, prepare them for school, and lay a strong foundation for lifelong learning.

At Too Small to Fail, we've found that one of the most effective ways to bolster early childhood brain development is to encourage meaningful interactions anywhere, anytime—in places like laundromats, playgrounds, pediatricians' offices, and anywhere you're changing a diaper. In 2017, for example, we partnered with Spotify to create playlists that also featured prompts and songs so that parents could sing with their children during bedtime, bath time, diaper time, and other everyday moments. We've also found that busy parents prefer receiving our tips and prompts via text messages instead of more traditional methods like email and snail mail. In fact, 90 percent of text messages sent in the United States are read within just three minutes of delivery. In 2015, Too Small to Fail partnered with Univision, the country's largest Spanish-language broadcaster, to launch a first-of-its-kind Spanish-language mobile messaging service to influence parent knowledge, attitudes, and behavior. To date, more than 125,000 subscribers have joined the service, which messages caregivers of children up to five years old with twice-weekly tips, prompts, and activity ideas to support parent-child engagement; boost early brain, language, and social-emotional development; promote bilingualism; and encourage early learning.

We're seeing concrete results from these initiatives. For example, we have seen that subscribers to the Univision/Too Small to Fail text-message service were more than twice as likely to talk, read, sing, and engage in simple math activities daily with their children. Perhaps most importantly, parents and caregivers who subscribe say they feel empowered, and 94 percent believe they personally can make "a great deal of difference" in helping their children succeed in kindergarten.

Other organizations are also powerfully using text-based communication to support parents and caregivers to be effective first teachers. Ready4K, ParentsTogether, Bright by Text, Text4Baby, and Vroom reach parents and caregivers via text and social media with activities, prompts, resources, and other tools to nurture parent-child relationships, improve health outcomes, and bolster children's early physical and emotional development.

While we must protect our children from technology in many ways, we also need to recognize the ways technology can help parents and caregivers protect their kids today and promote their healthy development over time.

What, Me Worry? *The Rise of Stealth Parenting*

JULIE LYTHCOTT-HAIMS

Author of the New York Times *bestselling book* How to Raise an Adult, *Julie Lythcott-Haims served as dean of freshmen at Stanford University for a decade.*

A mother loses sight of her young daughter at a park. For a few terrifying moments she fears the worst, but soon all is well. A new product offers Mom the reassurance she seeks—a chip that can be implanted in her daughter's brain with a corresponding app that lets Mom monitor her daughter's whereabouts and erase any distressing sounds or imagery in the environment.

Although the daughter is kept "safe," she is emotionally immature and becomes a social outcast. A psychologist persuades Mom to stop using the app (the chip cannot be removed). But as the daughter ages and starts to engage in risky teenage behaviors, Mom feels the need to monitor and manipulate her daughter's life once again. When the now sixteen-year-old daughter catches on to what Mom's up to, she grabs Mom's tablet and beats her unconscious with it. Then she flags down a truck driver and hops a ride out of town.

This is a scene from a 2017 episode of *Black Mirror*, the British anthology TV series. Set in a near-future world, it illustrates how technology alters human behavior. In this episode, "Arkangel," director Jodie Foster portends a futuristic version of helicopter parenting that may be closer at hand than we realize.

Helicopter parenting comes in the form of the worrywart who needs to constantly touch base with their child; the concierge who handles most transactional aspects of their child's life; and the authoritarian who mandates what their child must do all the time. This new parenting paradigm was first identified in 1990, and it went on to radically change childhood in middle- and upper-middle-class communities in the United States, Britain, and elsewhere. Psychologists report that these behaviors stem from a parent's personal insecurities, overblown fears, and need for control.

Facing mounting criticism in the last decade and a half, helicopter parents protest that they simply love their children and want the best for them, and they point to what I call short-term "gains": Under a parent's close watch, children *are* kept safer, make fewer mistakes, are shielded from unpleasant outcomes, and meet parents' expectations for their behavior and accomplishments. From that angle, it sounds like a recipe for success. Yet the long-term pains from helicopter parenting became evident when the first wave of children raised this way finished high school and began to leave home.

The first children subjected to the first playdates in 1984 became the first college students in the late 1990s whose parents couldn't let go and the first members of a generation deemed by employers and the media as "failing to launch." Their parents didn't let go, in part, because they had always overhelped—*so why stop now, in college, when the stakes are even higher*, their thinking went. But their children *were* in fact less capable than typical young adults of the same age because they had been so overhelped in childhood.

Beyond having underdeveloped life skills and workplace readiness, such young adults tend to have higher rates of anxiety and depression, are more likely to be medicated for anxiety and depression, and have greater rates of separation anxiety. Point being, when it comes to psychological wellness, it turns out that there *are* things even worse than not having everything go right for you. If someone does everything for you—instead of letting you try to do things for yourself and learn from your mistakes—they literally deprive you of a chance to develop a healthy self. This is called self-efficacy, and it's in short supply in overparented children.

Contrary to what some pundits and psychologists suggest, the smartphone didn't *cause* helicopter parenting. It was already in full swing at the college level in 2007, which is when the smartphone arrived. But the smartphone is a technological "enabler." Referred to by researchers as "the world's longest umbilical cord," it allowed parents of college students to start texting their "children" throughout the day—to check up on what they were doing, how they were faring, and what was happening next. By the late 2010s, college—which used to be characterized by freedom and independence—was now interruptible by parent texts and phone calls. Students were just as likely to be talking with their parents as with their friends before and after class. Most alarming to me was that the students themselves didn't seem to mind.

Closeness between parents and kids is of course a good thing—kids (and grown-ups) need human connection almost more than anything else. To be sure, we parents are physically closer to our kids than ever, reluctant to ever let them out of our sight, and more aware of their goings-on. Isn't that a good thing? It depends.

Train a more focused lens on connections between parents and kids today. Drive by your local elementary school pickup line, stand on the sidelines after a kid's soccer game, go into a restaurant at dinnertime, watch a kid doing homework at the dining table, and you're likely to see parents there, focused on a device. Is that connection? Listen to their conversations. They're likely to be about how their kid did on a test, whether they've started their homework and how much of it they have, and what they can do to run faster, swing harder, score more goals, or get higher grades. Missing are open-ended questions about how their kid is doing or feeling, and conversations about nothing at all or about life. As the author Jessica Lahey's work on intrinsic motivation shows, kids need autonomy as well as connection—room to think, space to try and fail and try again, and time to be by themselves for a while to get things done on their own.

Our obsession with knowing every detail about our kid and capturing and sharing it with others might be encroaching on our kid's need for autonomy. It might be more about our own insecurities than what's right for our kids. I can't help it: *I feel for these children.* I'm rooting for Gen Zers to develop an app that allows them to take the entire public record their parents created *about* them and wipe it clean.

Back in the day, a helicopter parent could affect only so many outcomes. But technology has marched forward, and with a fingertip and a screen, they can know where their kids are, see what they're doing, instantly know how they performed, and jump on it. Tech companies are profiting from parents' fears, and products arrive on the market before psychologists can say whether they're safe for kids.

Exhibit A is the "parent grades portal." At the start of the last decade, some software genius down the road from me in Silicon Valley introduced a feature into the K–12 schooling landscape that gives parents unlimited access to a teacher's gradebook. Many parents report checking it obsessively throughout the day. Often they learn of test results before their kid does. Instead of coming home to a snack, kids now come home to an inquisition: "You got a C on the science test? I thought we studied for that!" School becomes about frequent assessment and judgment instead of learning.

Then came location-tracking apps, which give parents peace of mind about their kid's whereabouts. "He missed his curfew but at least I know where he is," says the sheepish mom. When I ask parents why they use this technology, they usually say that the world is scarier and less safe today. Yet that's patently untrue; the rates of all types of violent crime have fallen steadily since the 1970s. Childhood was *less* safe when we were young, and none of us needed GPS tracking to stay alive. What *has* increased, however, is parents' fear that their kids are in peril if they aren't constantly monitored.

Then there's the ubiquitous web cam, which lets parents monitor the yard, front door, and every room inside the house. A guy sitting next to me on a plane watched his family come and go through the living room and got annoyed when one of his kids left a pair of shoes behind. "Pick up your shoes," he barked from thirty-five thousand feet, presumably through a virtual-assistant device. I don't know how the kid felt, but it sure creeped me out.

Taken together, these technologies have birthed what I call the "Stealth Parent," who uses surveillance to assuage their fears and exert their need for control. Kids are effectively monitored 24/7/365, whether at home, in school, or out in the world without us. Being subjected to constant monitoring—a fate once reserved for those who were a threat to themselves or others, such as incarcerated persons and patients in psych wards—has become life as usual for many children. Gizmo now makes a "wearable" device for young children that lets parents call them for dinner and monitor their whereabouts. How many years will it be before some company offers an implanted chip?

Helicopter parents have existed for over three decades. Anxiety and depression are on a steady rise in both children and adults.

Now comes surveillance technology and Stealth Parenting. Are they making kids and adults safer or more afraid?

It may be more expedient to let a chip inside a diaper tell us that our kid needs changing, but shouldn't our own senses alert us? Does it enhance or inhibit our connection when we yell at a kid from an airplane? Is our kid learning any lessons when they miss curfew, but we don't mind as long as we can see their blinking GPS dot? Is obsessively hitting refresh on the K–12 grades portal helping or hurting our relationship with our kid and their teachers?

Stealth Parents outsource to technology the one thing that we are uniquely positioned to do: parent our young. Inside our kids' developing brains—where we're supposed to be instilling love, trust, and connection—we're sending the message: *I don't think you can be successful without me.* This undermines their chance to build self-efficacy, a fundamental tenet of the human psyche. Without it, anxiety and depression bloom.

When I was a college dean, the majority of helicoptered students did not seem to mind all the help their parents gave them. But as the years went on, they began to grapple with what they could and could not do for themselves, their ability to cope with what life threw their way, and their continuing dependence. It began to dawn on them that something was amiss. One late twentysomething told me that, even though he was now in law school, his mother called him three times a day to check up on him. One day, he snapped. "Your voice is the only voice in my head," he screamed at her. "I need to hear my own voice!" He threw the phone across the room, and it was years before he spoke to her again.

I envision a future day when psychologists publish longitudinal studies correlating psychosis with being surveilled throughout childhood. If at some point, as contemplated by *Black Mirror*'s "Arkangel," the kids stand up for themselves and take matters into their own hands—forgive me, but I'll be rooting for the kids.

Is This the Culture That We Want?

JENNIFER SIEBEL NEWSOM

The First Partner of California, Jennifer Siebel Newsom is an advocate, founder of the Representation Project, and documentary filmmaker who wrote, directed, and produced the films Miss Representation, The Great American Lie, *and* The Mask You Live In. *Her husband, Gavin Newsom, is the fortieth governor of California.*

I remember one morning when my now ten-year-old daughter, Montana, was four and a half, and my husband, Gavin, turned to her and said, "Montana, do you think you'd ever want to be president?" Montana looked at us with her big, impressionable eyes and immediately responded with an answer that still shocks me to this day—"Oh no, Mommy and Daddy, girls can't be president. Only boys can be president."

You should have seen the looks on our faces. I was aghast. Gavin's jaw dropped. Here was our daughter telling us that women could not be presidents, as if it were the most obvious truth! We corrected her right on the spot, explaining that just because our president was a man and her books only showed pictures of past male presidents, it didn't mean that women *couldn't* be president. It just hadn't happened yet.

In America, your gender, unfortunately, matters—and these limiting narratives are so pervasive, so omnipresent, that they affect us from the moment we're born, even if we grow up in the most progressive of households. And they really do start with conditioning and our learned environments. As I explored in my 2011 documentary *Miss Representation*, our media—from TV and movies to YouTube, video games, and social media—continue to sell the idea that girls' and women's value lies predominantly in youth, beauty, and sexuality, not in the capacity to lead. Boys, meanwhile, learn that their success is tied to dominance, power, and aggression. In fact, all the way back to the womb, boys are socialized to believe that *anything* associated with femininity is inferior. They are encouraged to compete, form hierarchies, and utilize aggression, regardless of the cost to themselves or society—something I explored in my 2015 documentary *The Mask You Live In*.

As Gavin and I realized with a shock that morning, gender stereotypes have become so normalized in our society that we are often not even conscious of them. Everywhere we look, the marketplace and media continue to reinforce an imbalanced culture that devalues women and girls (*and* the traits we associate with them, like care and empathy), while encouraging dominance and aggression in boys and men. And it isn't just gender stereotypes that media and technology perpetuate. Too many of us, especially our children, are constantly absorbing a barrage of harmful messages, from racism to the normalization of othering and bullying. The toxic norms, imbalances, and hierarchies we *see* all across our media landscape underlie the larger, systemic inequalities we *experience* all around us. As a result, with our heads down and our eyes transfixed on our devices, we are in a "crisis of connection" where we aren't learning to connect with one another, listen to one another, and build healthy relationships and healthy communities.

In 2016, we got our society-wide wake-up call. We elected a president who came to fame through reality TV, which celebrates stereotypes and appeals to our basest instincts. Of course, President Donald

Trump didn't create the "crisis of connection." He is a symptom, a giant mirror reflecting back to us the distorted values of a culture that *we* created. Yet that ominous reflection has a silver lining. We can now ask with much more clarity: Is this the society we want to be living in? Are these the values we hold dear? Perhaps most importantly, is this the culture we want to hand down to our children?

As a working mother of four and the first "first partner" of California, I feel a great responsibility to ensure that the answer is a resounding *no.* Now, more than ever, we need to build a culture that uplifts us all, starting with the biggest cultural communicators of our time—technology and all forms of media. I do believe that is possible, if we make changes at the individual, corporate, and government levels.

As individuals, we need to start by experiencing life beyond our screens. We need to set technology and media limits with our kids and use social media sparingly, for positive purposes. We need to teach our kids to move, to go outside and explore the many wonders of the world. We need to teach them to be engaged citizens and participate in creating culture, since this work will not end with us. We must also break bread with people who appear different from us—look one another in the eye, have real conversations, and search for common ground.

At the corporate level, media and tech companies must no longer ignore their responsibility to the common good. They must care about multiple bottom lines—not only *how well* they have lined their pocketbooks (and those of their shareholders and investors), but also by *how much* they have contributed to the health and well-being of the communities around them, especially our most vulnerable citizens, our children.

At the government level, our leaders need to start asking hard questions about the role of media and technology in our lives and making tough decisions. If they don't, we need to vote them out because in the age of false information, rising divisive rhetoric, and twenty-four-hour infotainment cycles, we need to do more to protect our impressionable youth, who will create the culture of the future.

There is still time to right this ship. As a documentary filmmaker, I believe wholeheartedly in the power of media to inspire us and transform society for the better—if we use it correctly. We created this culture, and collectively, we can *re-create* it. Our children depend on it. They have so much promise and potential. Let's give them the tomorrow that they deserve.

PART 3

A THREAT TO DEMOCRACY?

Bolstering Democracy's Immune System

CRAIG NEWMARK

Craig Newmark is the founder of craigslist and Craig Newmark Philanthropies. Through his foundation, he helps advance grassroots organizations that are "getting stuff done" in areas that include trustworthy journalism and the information ecosystem, voter protection, gender diversity in technology, and veterans and military families.

Information warfare poses serious threats to our democracy. If we don't know who or what to trust in the news, we can't make informed voting decisions or otherwise participate in civil society. To really tackle the issue, we need collaboration from technology, news, security, academia, watchdogs, and others.

Since 2015, bad actors have been weaponizing information technology against the United States and other Western democracies to sway public opinion and interfere in our elections. These adversaries use botnets and exploit unsuspecting social media users to coordinate attacks, working around the clock to amplify and spread disinformation. During the 2016 US presidential election, fake news and conspiracy sites published more than six million tweets in October alone. Much more recently, in November 2019, a network of thousands of automated Twitter accounts targeted Kentucky voters with a disinformation campaign on election night, claiming that the local gubernatorial race had been rigged.

In response to these types of attacks, Facebook has taken down billions of fake accounts, but sophisticated propagandists endlessly find ways to work around their detection traps.

At the same time, our civic infrastructure is at risk. Outdated voting technology and technology that does not leave a paper trail is subject to compromise. Plus, those who help facilitate and improve the election process are potential targets of cyberattack.

While adversaries wage an information war on us and threaten to tamper with our systems, domestic bad actors are suppressing voters in twenty-five states, further jeopardizing the integrity of our elections. Unjust voter ID laws, gerrymandering, voter purges, and other restrictions make it so that millions of Americans are discouraged, harassed, and even forbidden from exercising their right.

That's the bad news, but there is hope. Americans are resilient, and once we get the message, we respond by working together.

There is no single solution to this crisis. Instead, we must support players from across the information ecosystem and make sure that they collaborate. It will take skilled folks in many different industries working together to figure out how to win this next stage in the information war.

One important step is understanding where the disinformation comes from and how it spreads, which includes ongoing investigations at the Stanford Internet Observatory. They are working to foster relationships with social media platforms so that they can leverage their user data to create a hub of information that researchers and nonprofits can then analyze. This will help us understand how false information spreads and develop strategies to weaken its impact.

We also need to help prevent newspapers from being gamed into doing the work of our adversaries and ensure that the press remains the "immune system of democracy." Adversaries work to get mainstream media—the source of news for most Americans—to amplify disinformation. Journalists need serious training in how to spot manipulated and falsified content, which is becoming more and more sophisticated. The global nonprofit First Draft and their CrossCheck initiative offer an online verification network that brings together newsrooms from around the world to identify and report false, misleading, and confusing claims that circulate online. Ahead of the 2020 US presidential election, the organization is offering free training and technological support to help build newsroom resilience to manipulation around the country.

To help ensure that newsrooms are not further implicated in the spreading of false information, they need to adjust their norms for ethical reporting. Dan Gillmor—cofounder and director of the News Co/Lab at Arizona State University's Walter Cronkite School of Journalism and Mass Communication—pointed out that some news outlets give considerable airtime and print space to people who they know are lying, thereby amplifying disinformation. He characterizes them as being "loudspeakers for liars" because these outlets end up repeating blatant deceptions for their audiences to absorb. To be clear: We are not talking half-truths or gray areas here, but the "pants on fire" stuff.

To avoid being complicit, news organizations will have to institute new reporting methods. Tactics may include adopting the "truth sandwich," which means covering a lie by presenting the truth first and then following the lie with a fact-check. They may also increase newsroom capacity to check claims for accuracy in real time, prior to publishing a story.

To that end, ongoing work at the Columbia Journalism School is strengthening the journalism ethics curriculum for students, while the Poynter Institute is conducting relevant training and continued education for professionals. These programs play critical roles in modernizing ethics so that the news industry keeps pace with the ever-changing digital landscape.

Equally important, we need to help restore a vanishing local press, which is critical to keeping the public informed and safeguarding our country from manipulation. With this goal in mind, the American Journalism Project is aiming to rebuild nonprofit local journalism through venture philanthropy. Complementing this effort, Report for America is placing hundreds of emerging journalists in newsrooms around the country to bolster reporting on local issues, especially those that have long gone underreported.

If the people in these efforts continue to work together, I have a lot of confidence in America's ability to defeat our foreign adversaries and their domestic allies. Good folks are pushing hard to make that happen, but it's all hands on deck, and everyone needs to stand up.

Reclaiming Democracy

MARIETJE SCHAAKE

As a former member of the European Parliament for the Dutch liberal democratic party, Marietje Schaake focused on trade, foreign affairs, and technology policies. She is currently president of the CyberPeace Institute, international policy director at the Stanford University Cyber Policy Center, and an international policy fellow at the Stanford Institute for Human-Centered Artificial Intelligence.

The technological revolution went hand in hand with the promise that technologies would bring about democracy. Company leaders, eager to be on the right side of history, explicitly argued as much. When in 2011 demonstrators' braved bullets in Cairo's Tahrir Square and faced torture in Egyptian police stations, headlines in Western media cheered on "Facebook and Twitter revolutions." Vodafone even claimed credit for the liberating force of popular youth movements in an advertisement. The promise of empowerment and greater agency through technological innovations was appealing, especially for people living under repression. But while mobile phones and internet access did allow people to mobilize, organize, and document human rights abuses, tech leaders grossly underestimated the power imbalance between individuals and repressive states.

Now we are learning, the hard way, that technology did not strengthen democratic governance. Corporations, not democratic governments, have actually usurped the task of governing through business models and technological standards. Without popular legitimacy or balancing majority and minority interests—without checks and balances or independent oversight—the power of governing online information ecosystems is now in the hands of gigantic tech companies. We are in the middle of a rude awakening and must urgently join forces to reclaim democracy.

What did these companies do to prevent digital spying on people through their products and networks and the explosion of viral hate speech in countries with known records of violence? If they actually wanted and intended to avoid harm and promote democracy, why didn't they prioritize these goals in product design and business decisions? Those decisions have now become serious challenges to the resilience of democracy, even in democratic societies.

As the power of technology platforms grew exponentially—before the Cambridge Analytica scandal—many claimed that their ability to steer election outcomes was imaginary. When European politicians like myself began to propose regulations to put principle over profits, tech lobbyists cautioned that any regulation would impede the liberating benefits of their well-intended enterprises. It has taken decades, and serious harms, for a wake-up call: Politics and democracy have not been strengthened but hijacked and dangerously disrupted.

A core element of tech's promise of democracy was that, by enabling greater access to information and disrupting top-down power structures, it would empower individuals versus monopolists. Instead, however, tech giants themselves are now the gatekeepers of data and information flows. They use power for their own purposes and profit, accountable only to their shareholders. Their promise of fostering transparency has brought us opaque algorithms, with scarce information to enable independent, democratic oversight. Instead of giving a voice to the voiceless, they have handed virtual megaphones to hatemongers and amplified their messages through unidentified bots. Public debates have become privatized. The largely unregulated platforms facilitate the spread of disinformation and polarization while driving up their ad revenues and time spent online. Data brokers profile people's most private details. Meanwhile, the erosion of trust and participation is pulling the threads of democracy's fabric. Propagandists gain superpower by microtargeting messages and ads. While democracies carefully balance majority and minority interests, tech giants steer toward maximum profits. Those who can pay or game their messages to the top of search results and feeds get maximum visibility, while other voices are remaining invisible and unheard.

There is no doubt that online manipulation can cause real-world harms. Murders have been inspired by online conspiracies. Such cases anecdotally demonstrate the power of digitally persuading a small number of people to change their behavior in the real world. With elections hinging on small differences, the ability to change the minds of a small portion of the population can have decisive effects on their outcomes. Amid deployments of coordinated inauthentic behavior campaigns, the very idea of freedom of choice comes under question.

The irony is that, while they lobby against regulation, the technology giants themselves have taken over more and more responsibilities that governments traditionally dealt with exclusively. Technology companies operate the foundation of entire information ecosystems for billions of people worldwide, from building and developing critical digital infrastructure to securing it and running search engines and social media platforms. They are the unrecognized, uncontrolled regulators of our time. Digitization almost always means privatization. With their deep and wide reach, corporations have become the governing actors of the online world. They develop standards and initiate rules or agreements, but without the mandates and accountability that democratic governments depend on to act legitimately. When the most powerful companies take over the initiative of global rule-making and cross-border governing, of setting and enforcing standards, they are directly challenging democracy and the rule of law.

Alibaba's cofounder Jack Ma makes an explicit case for corporations taking over the role of governments: "Innovation always develops much faster, and I think future laws should not be driven only by governments; they should be driven by private sector and all stakeholders together." Yet democracy, as a multistakeholder, deliberative process, is not designed for speed. It is designed for principle.

While technology companies have enhanced their power, democratic governments have moved in the opposite direction, taking a remarkable step back. Their reluctance to extend an international rules-based order to the online world is a historic break with the position that the United States and European countries have taken since World War II. From trade and human rights to rules that apply in wartime, democratic governments painstakingly built support for an international rules-based order after the devastation of World War II.

That leadership stopped with the technological revolution. Tech giants thus far deny researchers, parliamentarians, watchdogs, and journalists meaningful access to information, hiding behind trade-secret protections. Without independent oversight of their business models and algorithms, we cannot assume their compliance and cooperation to support democracy and respect for the rule of law. Ending up on the right side of history requires governing for principles, independent oversight, and enforcement in cases of violation of rules.

To reclaim democracy in the public interest, access to information about the inner workings of algorithms is essential. From accountability of AI and preventing election meddling to the proliferation of cyberweapons, we need to strengthen democratic governance of technologies on the local, national, and multilateral level. We must ensure that laws apply online as they do offline. The United Nations and European Union have stated this as their objective, but to achieve it, we need to empower regulators with skills and mandates. Discrimination watchdogs should assess whether algorithms are discriminatory, and if they are, consequences should follow, as they do in the real world. Data-protection regulators and antitrust authorities will need the knowledge and capacity to assess data sets and weigh whether their processing compromises tech companies' obligations to comply with existing laws. By putting principles at the heart of regulations and empowering mandated institutions to assess violations, we will not need to pass new laws for every new technological advance.

Global guidelines, treaties, and democratic models will also have to be crafted for the application of universal human rights, digital trade flows, and the norms of responsible behavior in peacetime. Given the challenges of multilateralism these days, we will likely see local and national initiatives before we reach global consensus.

Will corporate leaders act against online manipulation, support accountability mechanisms, democratize the developments of their systems by giving users and impacted communities a voice, and assess unintended negative consequences? Will they push for democratic regulations, beyond those selected to strengthen their already firm market dominance? Will they continue to sell products and services in repressive countries with known track records of violating human rights?

Going forward, as the pressure on democracy rises globally, corporate leaders will have to articulate the values and principles they stand for. If companies want to play government, even temporarily, they must expect to be held to account, like governments are.

A combination of hubris and ignorance led to the promise that the technological revolution would help spread democracy across the world. Instead, nondemocratic forces have been empowered to hijack democracy by exploiting the vulnerabilities of unregulated connectivity. Let us avoid these same mistakes going forward and choose unequivocally to ensure that democracy is not further disrupted. The right side of history is with democracy and strengthening its reach and resilience.

Using Technology to Defeat Democracy

LATOSHA BROWN

LaTosha Brown is cofounder of the Black Voters Matter Fund—a power-building, Southern-based civic engagement organization—and a fellow of the Harvard Kennedy School Institute of Politics.

Technology is agnostic until it's designed not to be. You can use technology in a way that expands opportunity or limits it. In this democracy, technology is being used not to expand participation in the vote, but to suppress it.

It's amazing that we trade and move trillions of dollars online in a seamless process, in a matter of seconds, all over the world, on many platforms. So, if we can move all that money so securely over the internet, it's simply a false idea that you can't have safe voting online. That's just another part of the narrative that tries to mystify the idea of voting, so that those who are not in favor of democracy can create fear in people. Clearly, the technology exists to have effective, fair, and inclusive online voting, as well as easier voter registration, ballot access, and centralized voting information. There are so many potential ways to use technology as a corrective tool to expand democracy.

Instead, we've seen the systematic use of technology to exclude people from the voting process. The state of Georgia, where I live, is one example. When you register to vote in Georgia, your signature has to exactly match the signature on your government-issued ID. If it doesn't match—even in a very minor way, like an added initial—the software kicks your registration out of the system. The software also rejects punctuation marks that are common in African American and Latinx names, including apostrophes, accent marks, hyphens, and tildes. As a result, the signatures of thousands of voters, especially Blacks and Latinos, do not match the spelling in the state's database, creating a major obstacle to exercising their right to vote.

Georgia has also been spending millions of dollars on new equipment that is supposed to make voting more efficient and effective. But we're seeing the opposite. During the June 2020 primary election, when people voted for their chosen candidate on one of the state's new touchscreen systems, the machine would sometimes incorrectly select a different candidate's name. I experienced that glitch myself. The poll workers had to reset my machine and give me another ballot so that I could start over again. But how many people didn't notice the error or weren't given the opportunity to vote again?

There are other ways that people are using technology to suppress the vote. For example, on voting day in Georgia, poll workers had to input a key code to start operating the new machines. But many poll workers said they were never given the key code, so voters weren't able to use those new machines. In other cases, poll workers said they were never given cords to plug the machines in. Human beings are responsible for that. People are using technology as an excuse to make it harder to vote.

If we have the technology to go into space and identify the elements of rocks on Mars, we have the technology to run effective elections. We could have same-day registration and same-day voting, and we could calculate votes just like we calculate money. We could use technology as a tool to strengthen and help further democracy.

But there's always been an element in this country that sees democracy as a threat to their positions of power. Technology may be agnostic, but it's doing exactly what the people who order, design, and implement it intend it to do.

The Assault on Civil Discourse and an Informed Electorate

SENATOR MARK WARNER

Senator Mark Warner was elected to the US Senate in November 2008 and reelected to a second term in November 2014. He serves on the Senate Finance, Banking, Budget, and Rules Committees as well as the Select Committee on Intelligence, where he is the vice chairman.

In 1791, James Madison asserted that "whatever facilitates a general intercourse of sentiments . . . is favorable to liberty." Today, we still hold that the Madisonian values of free expression and a robust, thoughtful public discourse are touchstones of the American democratic experiment. Unfortunately, like many of our institutions today, that discourse is under assault from dark money, Russian trolls, and extremists who exploit social media. In each of these cases and others, we see openness, diversity, and accountability in the public sphere being undermined.

The internet optimism of the 1990s and 2000s obscured this trend for many of us. For years, a bipartisan consensus felt that the internet was inherently democratizing and liberating. We based major policies on this belief—from Section 230 of the Communications Decency Act to our stance on trade relations with China to the peaceful uprisings in the former Soviet Union. As a result, we now face serious policy challenges stemming from our naïve optimism.

Today, I would argue that we need to rethink this perspective and develop a new set of foreign and domestic tech policies that are based on a more pessimistic, or at least more *realistic*, notion of technology and the internet.

This is a somewhat surprising position for me. For many years, I've been the "tech guy" in Congress. Like most policymakers, I shared the consensus view that new technologies and the companies that built them were largely positive forces. But now we see how the misuse of technology threatens our democratic systems, the robustness of our economy, and our national security.

Russia's attack on our democracy awakened a lot of people to this truth. We know the United States faces serious threats in the cyber domain from both state and nonstate actors—not to mention the threat of misinformation and disinformation efforts by Russia and those who have copied their playbook. As a result of this recognition, we are finally beginning to have some overdue conversations about privacy, data transparency, competition, and other critical issues related to social media. We must also confront the ways that domestic actors have exploited these technologies.

Our position as a global leader on technology issues has been weakened by the retreat of the United States on the global stage, as well as by Congress's unwillingness or inability to formulate smart policy responses to the technology challenges we face. Frankly, I worry that the haphazard approach to trade we see today may end up exporting and internationalizing some of our worst technology policies. While I am encouraged that governments around the world, including the European Union, have begun to fill this vacuum, the need for US leadership on pragmatic, tech-savvy policy has never been greater.

The Russian Wake-up Call

As vice chairman of the Senate Intelligence Committee, I spent the better part of two and a half years following the 2016 election on the only bipartisan investigation into Russia's attack on our democracy. The bottom line is, the United States was unprepared in 2016, and both policymakers and our social media companies failed to anticipate how platforms could be manipulated and misused by foreign operatives. Quite honestly, we should have seen it coming. For one, many of the techniques relied upon by the Russian Internet Research Agency (IRA) were not new. The audience-building, the exploitation of recommendation algorithms, and other techniques were longstanding tactics of online fraudsters. We even saw an early warning sign in the context of something called "Gamergate" in 2014. Gamergate was a concerted harassment campaign waged against women in the video game industry. It foreshadowed how bad actors could use a variety of online platforms to spread conspiracy theories and weaponize Facebook groups and other seemingly innocuous online communities.

We also missed the warning signs in the international context. Over the last two decades, adversary nations like Russia have developed a radically different conception of information security—one that spans cyberwarfare and information operations. I fear that we have entered a new era of nation-state conflict: one in which a nation projects strength less through traditional military hardware and more through cyber and information warfare. Already we have seen this Russian playbook expand to other nations: Social media disinformation has been used by the Burmese military to instigate ethnic cleansing; by China to attack prodemocracy demonstrators in Hong Kong; by leaders in the Philippines to silence critics; and by shadowy digital marketing companies to prop up the military in Sudan.

In many ways, we brought this on ourselves. We live in a society that is becoming more and more dependent on technology products and networks. Yet the level of security and integrity we accept in commercial technology products is shockingly low. As a society, we continue to have entirely too much trust in the technologies our adversaries have begun to exploit—and in the capacity and wherewithal of technology providers to anticipate, identify, and eliminate these practices.

Working Together: An International Approach

While some in the private sector have begun to grapple with these challenges, many more remain resistant to the changes and regulations needed. Certainly, Congress has not had its act together, either. Unfortunately, it's not enough to simply improve the security and integrity of our own infrastructure, computer systems, and data. We must work in a coordinated way to deal with adversaries and bad actors who use technologies to attack our democratic institutions. It is imperative that we develop new international rules and norms for the use of cyber and information operations. We also need better mechanisms to enforce existing rules.

But norms on traditional cyberattacks alone are not enough. We also need to bring information operations, like those perpetrated by Russia in 2016, into the debate. In addition, we need to build international support for rules that address the internet's potential for censorship and repression. We need to present our own alternatives that explicitly embrace a free and open internet. And we need that responsibility to extend not only to government but to the private sector as well. The truth is: Western companies that help authoritarian regimes build censored apps or walled-garden versions of the internet are just as big a threat to a free and open internet as government actors.

More broadly, we need to realize that the status quo just isn't working. For over two decades, the United States has maintained and promoted a completely hands-off approach to internet governance. Today, the large technology platforms are, in effect, the only major part of our economy without a sector-specific regulator. For years, we told the world that any tweaks around the edges would undermine innovation or create a slippery slope to a dystopian internet. Instead, the opposite has happened: Many countries have gravitated toward a dystopian Chinese model, in part because we have not offered a pragmatic, values-based alternative—a vision of internet governance that enshrines values of free expression, pluralism, and privacy, while devising pragmatic rules of the road that prevent abuse, harassment, the proliferation of misinformation, and consumer harms.

It is evident that laws originally intended to promote good behavior—like Section 230, which protects platforms from liability and was meant to incentivize effective moderation—are used by the largest platforms as a shield to do nothing. In 2018, Americans were defrauded on Facebook to the tune of $360 million by identity thieves posing as military service members. The perpetrators aren't sophisticated state actors using fancy masking tools. They're fairly basic scammers in internet cafes in West Africa. But Facebook faces no meaningful pressure to do anything about it. Neither the defrauded Americans nor the impersonated service members can sue Facebook because of Section 230. To make matters worse, platforms like Facebook face little, if any, real competition. While Section 230 was born out of an era with more vibrant competition on the web, it rests on the now-incorrect assumption that sites would pursue robust moderation because users would flock to other providers if these sites became dangerous or abusive places.

This is just one example of the internet governance regime we've convinced ourselves—and tried to convince the rest of the world—is working fine. Instead of dealing with the misuse of their platforms, these large companies have externalized the responsibility of identifying harmful and illegal activity to journalists, academics, and independent activists. Rather than promoting pragmatic rules of the road for the digital economy, the United States continues to promote a laissez-faire approach to technology—whether that's refusing to sign the Christchurch call to action to eliminate violent or terrorist content online or continuing to include new platform safe harbors, like Section 230, in our trade agreements.

As Americans, we must begin a more robust debate that acknowledges that the facts on the ground have changed since the 1990s. I have put forward a white paper and several bills that lay out a number of policy proposals for addressing these challenges.

We can start with greater transparency. Most Americans can agree that users have the right to know if the information they're receiving is coming from a human being or a bot. Companies should also have a duty to identify inauthentic accounts. If someone says they're Mark from Alexandria but they're actually Boris in St. Petersburg, people have a right to know that. If a large Facebook group claims to be about Texas pride, but its administrators are consistently logging in from Moldovan and Belarussian IP addresses, the users following that group's posts should know that, too.

We also need to put in place some consequences for social media platforms that continue to propagate truly defamatory content. In 2019, Facebook was caught flat-footed in the face of rudimentary audiovisual manipulation of a video of Speaker Nancy Pelosi that attempted to make her appear drunk. This does not bode well for how social media is going to deal with more sophisticated manipulation techniques like highly realistic deepfake videos. Platforms should be granting greater access to academics and other analysts studying social trends like disinformation to identify problems—before they reach a scale that threatens our democracy. Instead, we've seen a number of cases where platforms have worked to shut down efforts by journalists and academics to track misuse and abuse of their products.

We can't stop there. We must work with social media platforms to continue broader conversations around privacy, price transparency, and interoperability. It's my hope that these companies will collaborate and be part of the solution, but one thing is clear: The Wild West days of social media are coming to an end. Democracies have been at the forefront of technological innovation. The postal system and the telegraph and the radio were all essential to the development of our republic; in their own ways, they cultivated a flourishing of informed discourse and an informed electorate. In order to continue this leadership and preserve our own civil discourse, we need to question outdated assumptions about today's communications technologies and put forward policy ideas that are consistent with our values.

Repairing a Fractured America

LAWRENCE LESSIG

Lawrence Lessig is the Roy L. Furman Professor of Law and Leadership at Harvard Law School and the former director of the Edmond J. Safra Center for Ethics at Harvard University. His most recent book is They Don't Represent Us.

In a nation dedicated to the freedom of the press, it's very hard—not to mention undesirable—to legislate limits on political speech. That cannot be the role of government if democracy is to remain free of state control. But as information channels have multiplied, a real "broadcast democracy"—a shared and broad engagement with a common set of facts—has disappeared.

The Civil War may well have been the last time we suffered a media environment like the present. Then, it was censorship laws that kept the truths of the North separated from the truths of the South. And though there was no polling, the ultimate support for the war, at least as manifested initially, demonstrated to each of those separated publics a depth of tribal commitment that was as profound and as tragic as any in our history. That commitment, driven by those different realities, led America into the bloodiest war in its history. We're not going to war today.

We are not separated by geography, and we're not going to take machetes to our neighbors. But the technical environment of media in our culture today leaves us less able to work through fundamental differences than at any time in our past. Indeed, as difference drives hate, that hate pays—both the media companies and too many politicians. But the cost to the republic of this business model of hate is profound. The nation needs at least temporary, if voluntary, restraint to help us through these ugly times. This is a moment to knit common understandings, not a time to craft even more perfectly separated realities.

That knitting could begin with both networks and digital platforms asking not what is best for them, individually, but what would be best for us all, together. Which network or platform strategies will enable a more common understanding among all of us? And which strategies will simply drive even more committed tribe-based ignorance? If we as a people are to be persistently polled and our views so persistently legible to our representatives, then at least we should know enough in common to make judgments in common.

Fox, MSNBC, and the others should push opinion-based reporting to the side and place journalism-based news in prime time. They all must take responsibility for their audience understanding the facts, more than simply rallying its side to its own partisan understanding. Partisan networks may not be a bad thing in general. They are certainly a bad thing in moments like this.

Social media platforms have responsibilities here as well. We don't yet know the consequences of those platforms forgoing political ads in the context of an entire election season—even as some experiment with doing so. But Facebook and Twitter together could take the lead in turning away ads aimed at rallying a base or trashing the opposition.

More fundamentally, platforms could block falsity better. Intellectual property on the internet has long been protected by a notice-and-takedown regime. If a platform gives copyright owners an easy way to notify it about copyright violations, and if it removes those violations quickly, then the platform is not liable for the infringement. It is time we extend a similar mechanism to defamatory speech. If a platform has been shown the falsity in what it continues to publish, its continued publication should be considered "actual malice," and thus no longer immune from liability. Platforms without editors cannot be immune from responsibility—especially when the incentives of clickbait become so central to the business model of online publishing.

None of this, of course, is likely to happen anytime soon. But we should not underestimate the potential for leadership. There is an equivalent to peaceful nonviolent protest—an act that so surprises the other side that it forces a recognition that otherwise would be missed. Any prominent actor in the midst of this mess who stepped above the common play might surprise enough to trigger a change. Here, too, could be a role for former presidents. Why don't we see George W. Bush and Barack Obama standing together on this, not by directing a result but by counseling a better process?

No doubt, all this is a big ask—lucrative networks and social media platforms unilaterally disarming or agreeing to a new set of rules. But there's another way to look at it. Businesses succeed by managing risk, and the risk of a truly destabilizing event—a fractured America because of siloed information—is much greater than the risk of losing some ratings for a few weeks or months.

There is no mechanism that guarantees a democracy's safety. There is only, and always, the courage of individuals to be better than anyone expects.

Technology for Global Good

KHALED HOSSEINI

Born in Kabul, Afghanistan, Khaled Hosseini is the author of The Kite Runner *and other novels. When he was fifteen, his family was granted asylum in the United States after the Soviet invasion of Afghanistan. He studied medicine and was a practicing physician from 1996 to 2004. Since 2006, he has served as a goodwill ambassador for the United Nations High Commissioner for Refugees—UNHCR, the UN Refugee Agency—and he is the founder of the Khaled Hosseini Foundation, a nonprofit organization that provides humanitarian assistance to women and children in his native Afghanistan.*

A couple of years ago, I spoke at the Tech Interactive museum in my hometown of San Jose, California. The topic was "tech for global good"—which is the name of the Tech museum's flagship program to help create the next generation of social innovators and change-makers who will tackle our planet's toughest challenges. A writer friend of mine was also asked to speak, but he was reluctant, due to his ever-deepening skepticism of technology. But when he arrived at the museum and saw the *Body Worlds* exhibit, the BioDesign Studio, and the Exploration Gallery, where kids can experience earthquakes and weightlessness like astronauts, he beamed and said to me, half-jokingly, "Oh, so you meant *good* tech. Why didn't you say so?"

This is the extent to which the bloom has come off the technology rose. Big Tech has become a pariah, as the list of societal ills it is blamed for continues to grow—from breaches of data privacy to the proliferation of fake news, from mass surveillance and cyberattacks to spawning a generation of youth addicted to clicks and likes. But perhaps Big Tech's biggest sin has been enabling the digital threat to democracy. Critics point to the flood of digital disinformation pivotal in the 2016 US elections—as well as in elections in the United Kingdom and France—and blame it, in part, for the wave of nationalism and populism that has inflamed political polarization and resentment. Big Tech, the charge goes, is ruining democracy.

But in other places, notably my birthplace of Afghanistan, the relationship between technology and democracy is playing out a little differently. Access to technology is enabling millions of Afghans, most of them young, to engage with the democratic process.

Afghanistan today is still, in many ways, a struggling country. Though millions of kids have returned to school and perennial scourges like infant mortality and maternal mortality have been curbed, the country has not yet achieved stability. Natural disasters, a poor economy, and inadequate access to social and health services have forced many Afghans from their villages. Millions of returning refugees have difficulties restarting their lives in rural areas, where only 12 percent of the land is arable. Countless Afghans have become internally displaced in the countryside due to insecurity. Indeed, Afghan civilians are being killed at record levels. According to the United Nations, the first nine months of 2019 saw more than eight thousand Afghans killed, most by insurgent suicide attacks or by IEDs.

Hoping to find elusive security and opportunity for their families, Afghans are flocking to the cities like never before. In Kabul, for instance, the population has ballooned from 1.5 million in 2001 to more than 4 million today, making it one of the fastest-growing and most-congested cities in the world. This shift has proven to be a huge challenge. Limited resources are being further strained.

Urban slums are on the rise. Living conditions are increasingly crowded and difficult. Unemployment runs high. Other cities in the country are also expanding and struggling to adapt to this rapid urbanization of a traditionally rural nation.

One of the by-products of this urbanization process has been an increased access to technology for millions of Afghans and, along with it, increased connection to the country's neighbors and the world at large. The high penetration of technology like cell phones (up to 85 percent of the country has mobile phone services today) has made it easier for entrepreneurs to open businesses, for students to study for college entrance exams, and for the government to manage natural disasters.

Technology has also opened doors for millions of young people. Over 60 percent of Afghanistan's population of 27.5 million is under the age of twenty-five. They are engaging with issues such as climate change, human rights, education, and participatory politics. Information technology has enabled Afghans to become more aware of social change and brought a people long cut off from world events by poverty and conflict face-to-face with wide-ranging issues. Afghans now enjoy a sense of connectedness to the outside world, as well as to their own world. Afghan internet users routinely debate the inherent complexities of Afghan politics, the role of Islam in a democratic state, ethnic issues, and the clash between conservative viewpoints and more progressive actors. What's more, Afghans are increasingly using technology to mobilize support and organize action on issues they feel strongly about.

The most moving example of this came in March 2018, when a grassroots peace movement began in Helmand and eventually came to be known as the Helmand Peace March. Protestors chanting slogans of "Stop the war" and "We want peace" traveled by foot through the battle-weary areas of Kandahar, Zabul, and

Ghazni provinces and marched into Kabul for a sit-in aimed to end the violence in the country. The movement struck a chord. News of its steady advance spread on social media. Young people who were invested in democracy and peace-building went online and created social media hashtags to support the cause and echo its demands for a cease-fire and resumption of peace talks. Social media activity became a sociopolitical mobilizer and recruited demonstrators and supportive voices in a way that would not be feasible through word of mouth.

Of course, no democracy can function without the active participation of women. Afghan women, a long-neglected group in the patriarchal country, are today directly benefiting from technology. Digital literacy is giving them, at least in urban areas, a chance at financial independence and a voice in the greater national conversation on issues that impact them. Thanks to the internet, Afghan women have a much wider vision of the world, as they become digital citizens in increasing numbers. Through technology, they are acquiring skills necessary for self-reliance and overcoming persistent gender-based limitations. While conditions for them are still unacceptably grim in many parts of the country, in urban settings the old paradigm is slowly shifting. Technology is spearheading a generational change among Afghan women and fostering positive social change to which, it must be repeated, persistent cultural obstacles remain.

This being Afghanistan, all of the above is highly precarious. Afghanistan faces an uncertain future. There is great anxiety over the eventual outcome of the stuttering peace talks and what it will mean to ordinary Afghans, especially the urban, educated, and digitally literate youth. The country continues to be wracked by poverty, violence, corruption, and the consequences of interference from its neighbors. Technology or no technology, no positive social change can persist without security and stability.

But the role of technology in the Afghan democratic experiment thus far has revealed a flip side to the current conversation about it in Europe and the United States: that of enabler versus inhibiter. That is not to say that technology will save Afghanistan, which is far from immune to fake news and digitally spread disinformation. Even established democracies are, as we have come to learn, fragile things. Afghanistan's nascent and troubled one is supremely vulnerable.

Still, when I think back to what my friend said about "good tech" (implying the existence of "bad tech"), I come back to square one: The only thing technology enables is human behavior. It amplifies both the good and the bad. Whether we are talking about democracy or humanitarian matters, technology itself changes nothing. We have, after all, the technological means to feed every human soul on the planet. Yet people are starving in Yemen, in South Sudan, in Nigeria, in Somalia. Technology, as the cliché goes, is neither good nor bad. It is instead like a boat. To be a force for greater good, it needs the wind of the right story blowing in its sails. Because, ultimately, it is those stories we tell ourselves, not our technologies, that free or oppress people, that erect or ruin economies, that preserve or hinder democracies.

The Era of Fake Video Begins

FRANKLIN FOER

Franklin Foer is a staff writer at the Atlantic *and teaches at Georgetown University. He is the author of* World Without Mind: The Existential Threat of Big Tech *and* How Soccer Explains the World: An Unlikely Theory of Globalization. *For seven years, he edited the* New Republic.

In a dank corner of the internet, it is possible to find actresses from *Game of Thrones* or *Harry Potter* engaged in all manner of sex acts. Or at least, the carnal figures look like those actresses, and the faces in the videos are indeed their own. Everything south of the neck, however, belongs to different women. An artificial intelligence has almost seamlessly stitched their familiar visages into pornographic scenes, one face swapped for another. The genre is one of the cruelest, most invasive forms of identity theft invented in the internet era. At the core of the cruelty is the acuity of the technology: A casual observer can't easily detect the hoax.

This development, which has been the subject of much handwringing in the tech press, started with the work of a programmer who goes by the nom de hack "deepfakes." And these videos are merely beta versions of much more ambitious projects. One of deepfakes's compatriots has boasted that he intends to democratize this work. He wants to refine the process, further automating it, which would allow anyone to transpose the disembodied head of a crush or an ex or a coworker into an extant pornographic clip with just a few simple steps. No technical knowledge would be required. Because academic and commercial labs are developing even more sophisticated tools for nonpornographic purposes—algorithms that map facial expressions and mimic voices with precision—the sordid fakes will soon acquire even greater verisimilitude.

The internet has always contained the seeds of postmodern hell. Mass manipulation, from clickbait to Russian bots to the addictive trickery that governs Facebook's News Feed, is the currency of the medium. It has always been a place where identity is terrifyingly slippery, where anonymity breeds coarseness and confusion, where crooks can filch the very contours of selfhood. In this respect, the rise of deepfakes is the culmination of the internet's history to date—and probably only a low-grade version of what's to come.

Vladimir Nabokov once wrote that *reality* is one of the few words that means nothing without quotation marks. He was sardonically making a basic point about relative perceptions: When you and I look at the same object, how do you *really know* that we see the same thing? Still, institutions (media, government, academia) have helped people coalesce around a consensus—rooted in a faith in reason and empiricism—about how to describe the world, albeit a fragile consensus that has been unraveling in recent years. Social media have helped bring on a new era, enabling individuated encounters with the news that confirm biases and sieve out contravening facts. President Trump further hastened the arrival of a world beyond truth, providing the imprimatur of the highest office to falsehood and conspiracy.

Soon this may seem an age of innocence. We'll shortly live in a world where our eyes routinely deceive us. Put differently, we're not so far from the collapse of reality.

We cling to reality today, crave it even. We still very much live in Abraham Zapruder's world. That is, we venerate the sort of raw footage exemplified by the 8-mm home movie of John F. Kennedy's assassination that the Dallas clothier captured by happenstance. Unedited video has acquired an outsize authority in our culture. That's because the public has developed a blinding, irrational cynicism toward reporting and other material that the media have handled and processed—an overreaction to a century of advertising, propaganda, and hyperbolic TV news. The essayist David Shields calls our voraciousness for the unvarnished "reality hunger."

Scandalous behavior stirs mass outrage most reliably when it is "caught on tape." Such video has played a decisive role in shaping the past several US presidential elections. In 2012, a bartender at a Florida fundraiser for Mitt Romney surreptitiously hit record on his camera while the candidate denounced "47 percent" of Americans—Obama supporters all—as enfeebled dependents of the federal government. A strong case can be made that this furtively captured clip doomed his chance of becoming president. The remarks almost certainly would not have registered with such force if they'd merely been scribbled down and written up by a reporter. The video—with its indirect camera angle and clink of ambient cutlery and waiters passing by with folded napkins—was far more potent. All of its trappings testified to its unassailable origins.

In 2016, Donald Trump, improbably, recovered from the release of a 2005 *Access Hollywood* tape, in which he bragged about sexually assaulting women, but that tape aroused the public's passions and conscience like nothing else in the presidential race. Video has likewise provided the proximate trigger for many other recent social conflagrations. In 2014, it took extended surveillance footage of the NFL running back Ray Rice dragging his unconscious wife from a hotel elevator to elicit a meaningful response to domestic violence from the league, despite a long history of abuse by players. Then there was the 2016 killing of Philando Castile by a Minnesota police officer. Castile's girlfriend streamed the aftermath of the shooting to Facebook, and all the reports in the world, all the overwhelming statistics and shattering anecdotes, would not have provoked as much outrage over police brutality as that video. The terrifying broadcast of Castile's animalistic demise in his Oldsmobile rumbled the public and led politicians, even a few hard-line conservative commentators, to finally acknowledge the sort of abuse they had long neglected.

That all takes us to the nub of the problem. It's natural to trust one's own senses, to believe what one sees—a hardwired tendency that the coming age of manipulated video will exploit. Consider recent flash points in what the University of Michigan's Aviv Ovadya calls the "infopocalypse"—and imagine just how much worse they would have been with manipulated video. Take Pizzagate—the conspiracy theory during the 2016 presidential election of a Democrat-run child-sex ring—and then add concocted footage of John Podesta leering at a child or worse. Falsehoods will suddenly acquire a whole new, explosive emotional intensity.

The problem isn't just the proliferation of falsehoods. Fabricated videos will create new and understandable suspicions about everything we watch. Politicians and publicists will exploit those doubts. When captured in a moment of wrongdoing, a culprit will simply declare the visual evidence a malicious concoction. President Trump, reportedly, pioneered this tactic: Even though he initially conceded the authenticity of the *Access Hollywood* video, he now privately casts doubt on whether the voice on the tape is his own.

In other words, manipulated video will ultimately destroy faith in our strongest remaining tether to the idea of common reality. As Ian Goodfellow, a scientist at Google, told *MIT Technology Review*, "It's been a little bit of a fluke, historically, that we're able to rely on videos as evidence that something really happened."

The collapse of reality isn't an unintended consequence of artificial intelligence. It's long been an objective—or at least a dalliance—of some of technology's most storied architects. In many ways, Silicon Valley's narrative begins in the early 1960s with the International Foundation for Advanced Study, not far from the legendary engineering labs clumped around Stanford. The foundation specialized in experiments with LSD. Some of the techies working in the neighborhood couldn't resist taking a mind-bending trip themselves, undoubtedly in the name of science. These developers wanted to create machines that could transform consciousness in much the same way that drugs did. Computers would also rip a hole in reality, leading humanity away from the quotidian, gray-flannel banality of *Leave It to Beaver* America and toward a far groovier, more holistic state of mind. Steve Jobs described LSD as "one of the two or three most important" experiences of his life.

Fake but realistic video clips are not the end point of the flight from reality that technologists would have us take. The apotheosis of this vision is virtual reality. VR's fundamental purpose is to create a comprehensive illusion of being in another place. With its goggles and gloves, it sets out to trick our senses and subvert our perceptions. Video games began the process of transporting players into an alternate world, injecting them into another narrative. But while games can be quite addictive, they aren't yet fully immersive. VR has the potential to more completely transport—we will see what our avatars see and feel what they feel. Several decades ago, after giving the nascent technology a try, the psychedelic pamphleteer Timothy Leary reportedly called it "the new LSD."

Life could be more interesting in virtual realities as the technology emerges from its infancy; the possibilities for creation might be extended and enhanced in wondrous ways. But if the hype around VR eventually pans out, then like the personal computer or social media, it will grow into a massive industry, intent on addicting consumers for the sake of its own profit, and possibly dominated by just one or two exceptionally powerful companies.

The ability to manipulate consumers will grow because VR definitionally creates confusion about what is real. Designers of VR have described some consumers as having such strong emotional responses to a terrifying experience that they rip off those chunky goggles to escape. Studies have already shown how VR can be used to influence the behavior of users after they return to the physical world, making them either more or less inclined to altruistic behaviors.

Researchers in Germany who have attempted to codify ethics for VR have warned that its "comprehensive character" introduces "opportunities for new and especially powerful forms of both mental and behavioral manipulation, especially when commercial, political, religious, or governmental interests are behind the creation and maintenance of the virtual worlds." As the VR pioneer Jaron Lanier writes in his recently published memoir, "Never has a medium been so potent for beauty and so vulnerable to creepiness. Virtual reality will test us. It will amplify our character more than other media ever have."

Perhaps society will find ways to cope with these changes. Maybe we'll learn the skepticism required to navigate them. Thus far, however, human beings have displayed a near-infinite susceptibility to getting duped and conned—falling easily into worlds congenial to their own beliefs or self-image, regardless of how eccentric or flat-out wrong those beliefs may be. Governments have been slow to respond to the social challenges that new technologies create, and they might rather avoid this one. The question of deciding what constitutes reality isn't just epistemological; it is political and would involve declaring certain deeply held beliefs specious.

Few individuals will have the time or perhaps the capacity to sort elaborate fabulation from truth. Our best hope may be outsourcing the problem, restoring cultural authority to trusted validators with training and knowledge: newspapers, universities. Even as big technology companies have shown belated awareness of this crisis—Facebook has banned deepfake videos—they have struggled to develop an effective response. Part of the challenge is that there's no simple algorithm that would allow Facebook to detect the fakes. It's perhaps telling that one of the most widely circulated fakes features a simulation of Mark Zuckerberg bragging that he has "total control of billions of people's stolen data."

Silicon Valley's failure to combat the problem may be as much ideological as it is technical. In 2016, as Russia used Facebook to influence the American presidential election, Elon Musk confessed his understanding of human life. He talked about a theory, derived from an Oxford philosopher, that is fashionable in his milieu. The idea holds that we're actually living in a computer simulation, as if we're already characters in a science-fiction movie or a video game. He told a conference, "The odds that we're in 'base reality' is one in billions." If the leaders of the industry that presides over our information and hopes to shape our future can't even concede the existence of reality, then we have little hope of salvaging it.

Adapted from "Reality's End," originally published in the Atlantic *(May 2018).*

The Informality Machine

YUVAL LEVIN

Yuval Levin is the editor of National Affairs *as well as the Beth and Ravenel Curry Chair in Public Policy and the director of Social, Cultural, and Constitutional Studies at the American Enterprise Institute. The author, most recently, of* A Time to Build, *he was a member of the White House domestic policy staff under President George W. Bush.*

When future generations look back upon the early decades of the twenty-first century, they will surely wonder at the bizarre spectacle of our society driving itself mad on social media. From high school students subjecting one another to horrific bullying online to our political and media elites abandoning all patience and decorum in an endless barrage of tantrums and counter-tantrums, from paranoid conspiracy theorists gaining traction and legitimacy to foreign governments picking at the rifts and scabs of our society, social media has exposed and exacerbated some grievous vulnerabilities.

On the face of things, we might expect social media to strengthen our institutions because the avowed purpose of social media platforms is to supplement our social lives: to connect people, to help us communicate and organize, and thus to help us do things more easily together. Social media has done all of that for many of us, without question. But as political scientist Joshua

Mitchell has noted, in practice social media platforms have often also become a substitute for our social lives, and as such they have done nearly the opposite of what we might hope: pulling us apart, encouraging aggression and hostility, and keeping us from hearing one another. Why is it that our social media platforms seem to be such a great fit for the vices and dysfunctions of our time, but such a poor fit for efforts to recover institutional integrity, civic peace, and solidarity?

This is particularly problematic as these platforms come to replace more of our experience of the broader world—to stand in for the formative experiences and connections of the public sphere. What happens on social media, in this sense, isn't quite public. It occurs in a space you design to your liking. Yet that space is not quite private, either, and it denies us some key benefits of private social interactions. In some respects, the online world seems overflowing with intimacy: Everyone shares very personal information—from pictures of their children to provocative opinions and jokes and complaints to extremely private dating profiles and intimate photos and messages.

But "personal" is actually very different from intimate or even private. In fact, the realm of social media often effectively functions as an arena for saying private things in public, and thereby confounding the public and the private in a way that renders social interactions deeply uneasy and unsatisfying. Some people, especially those who have grown up in the digital era, are sufficiently accustomed to this confusion that they do not even recognize it. They behave online as they might among their closest friends, even if a much broader circle of people can see them. This can have serious consequences—ranging from personal embarrassment to professional ruin or worse. Even people who seem perfectly at home with this peculiar ambiguity are often left starved for social intimacy and denied the opportunity to really let their guard down.

This confusion of public and private (rather than the kinds of worries we often hear about surveillance or the exposure of personal information) is the real privacy crisis of the internet age. It is a defining feature of social media and of life in our era more generally. Although we often cannot tell whether the world of social media is public or private, we can certainly call what happens in that world *informal.* We might even think of social media as a massive informality machine, robbing our interactions of structure and of boundaries. This is why moving more of our social activities onto the platforms of social media tends to bring the most dramatic and fundamental changes to those of our social interactions that would otherwise be most formal—like the presentation of professional work products, the intricate dance of dating and courtship, or the pronouncements of public policy. It is also why social media is uniquely corrosive of institutions, which are, after all, precisely social forms.

Formality has a bad name in our relentlessly democratic culture—we tend to equate it with stuffiness and rigidity. Informality, on the other hand, is synonymous with authenticity. But another way of understanding formality might be as a means of fitting social form to social function. It is a way of behaving when something important is at stake, which sends a signal regarding that importance, establishes a framework for its integrity and structure, and lends credence and protection to all involved.

From the formal vote that indicates a decision at a meeting to the letterhead that signifies the authority of an official notice to the structure of a scientific claim, formalities distinguish the exercise of legitimate power. They send us vital signals about what to take seriously and what to take lightly, when to speak and when to listen, who to trust and who to question. They offer us an architecture of behavior that makes it possible to have some predictability and security in high-stakes situations so we might successfully navigate the social world.

Online, all of these forms are abandoned. Everyone is just an individual on a platform. The results can be great fun but also very dangerous and confusing. The informality makes it hard for us to tell different kinds of expression apart, to judge what to believe, and to know how to behave. People often cross boundaries they would never think of crossing in an even slightly formal interaction in the offline world. Online, we frequently feel the painful absence of those barriers and boundaries.

The formalities of our social relations often mediate between individuals for their protection. They ensure, for example, that anonymity cannot be combined with proximity. If you communicate with someone directly, they know who is speaking to them; if you communicate anonymously, there will be some distance between the two of you. But on social media and in the online world, the absence of formal structures of mediation often translates to anonymity combined with proximity—and that combination can be perilous and toxic.

Anyone with even a modest public profile today can relate harrowing stories of nasty and pernicious harassment by anonymous trolls. Whether on Twitter (which allows for essentially anonymous profiles), in online comments on published writing, or by email, the internet often serves as a kind of open sewer pipe, spewing streams of excrement upon anyone who expresses an opinion in public. Anonymity relieves internet trolls of any need for the restraint that might be called for in the real world—out of concern for one's reputation, for example, or just from the healthy discomfort involved in treating someone else horribly. This is surely why people behave online in ways they would never contemplate in their offline lives.

In the course of the 2016 election, for instance, I found myself on the receiving end of intense and ugly anti-Semitism from various critics of my political writing, and I discovered that essentially every Jewish writer I knew could say the same. Others face far worse. Women with public profiles seem to suffer the most heinous abuses, often including abject threats of horrendous sexual violence—almost always anonymously.

Anonymity is not the only way that informality creates real problems on social media. In many cases, people who are not anonymous at all behave in online exchanges in ways they never would in face-to-face conversations. Many of the elites who shape our culture, economy, and politics have allowed themselves in recent years to be plucked out of the various institutions that normally refine and structure their work and to be plopped instead as individuals, unconstrained and unprotected, onto the exposed platforms of social media. Indeed, they have rushed to do this to themselves.

The result is a loss of restraint and protection—both of which are made possible by robust institutions. As individuals exposed on the platforms, we are always at risk of being betrayed by our own impulses and becoming ridiculous. The discipline and reticence so essential to leadership, professionalism, responsibility, decency, and maturity are forcefully discouraged by the incentives of the online world. Yet we all seem to find those incentives impossible to resist.

All of this results in a blurring of boundaries. Not only are the boundaries between public and private effectively erased in the realm of social media; where politics ends and entertainment begins is also increasingly difficult to say. The lines between activism, journalism, and conspiracy mongering are increasingly blurry; the boundary between academic research and political argument is, too. The ethos of reality television (in which it is never quite clear whether we are watching people live their real lives or act for our amusement) has come to dominate our experience.

"Is this real or is it for show?" That question has long hung like a cloud over celebrity culture. We have always had a vague sense that what we see of the lives of movie stars and other cultural icons is somehow choreographed. The same is true now of our politicians—who seem increasingly willing to be seen reading a script or playing a role with a half wink to the crowd. Politics practiced this way is utterly corrosive of the ethos of republican government. If people involved in the political world play roles in the sense that actors do, it becomes nearly impossible for them to play roles in the sense that leaders and citizens do. The same can be said of many other institutions. The ambiguity of celebrity theatrics now confounds vast swaths of our culture.

In fact, the specter of celebrity culture is a very useful lens through which to think about what happens on social media. By letting us carefully curate the image of ourselves available to others, social media encourages us to think of ourselves as living performatively. As we are experiencing an important milestone in our lives, or just enjoying a beautiful day with family or friends, a great many of us find ourselves thinking about how to capture this moment for Instagram or Facebook. As we undergo some mass experience of tragedy or joy or surprise, we work to convey an appealing image of ourselves experiencing it. Any time we are confronted with frustration—a delayed flight, an idiot boss, bad service somewhere—we yearn for a platform to vent our irritation to the world. When we find ourselves in some particularly lovely or notable place, or in the company of someone our circle of friends might know, we reach for the phone not to capture that person or place but to capture ourselves in proximity to them. The selfie culture is a culture of personalized microcelebrity, in which we each act as our own paparazzi, relentlessly trading in our own privacy for attention and affirmation and turning every moment into a show. As a result, we sometimes find it hard to really feel like we are living our lives unless we know others are watching us live them.

It may be, of course, that our society will develop novel antibodies to this set of problems in the coming years. After all, the internet and social media have not been around that long—and we are only now learning to live with them. Just as the first generations of human beings who lived packed together in cities faced all manner of new social difficulties, so we early denizens of the online world find ourselves confronted with bizarre and unexpected maladies. Over time, as we recognize these problems, we may find ways to live with them and to enjoy the benefits of information technologies while minimizing some of these costs.

Adapted from A Time to Build: From Family and Community to Congress and the Campus, How Recommitting to Our Institutions Can Revive the American Dream *(New York: Basic Books, 2020).*

PART 4

WHERE BIG TECH WENT WRONG

The Thief in Our Pockets: The Dark Side of Smart Tech

LAURIE SANTOS

Laurie Santos is a professor of psychology, head of Silliman College at Yale University, and host of The Happiness Lab *podcast.*

It's January 9, 2007, and Steve Jobs is addressing a huge crowd at the MacWorld trade show conference. After an update about Apple's ongoing work with Intel processors and Apple TV, Jobs looks away from the crowd and pauses. He takes a long sip of water and seems almost as though he is bracing himself for something big. "Every once in a while," he softly tells the crowd, "a revolutionary product comes along that changes everything. . . . Today, Apple is going to reinvent the phone, and here it is."

That fateful afternoon, Apple introduced the world to the first iPhone—an entirely new kind of invention that combined mobile phone technology with a pocket-sized internet communications device. The iPhone also had the same kind of operating system and software as the best Mac computers, all the amazing music capabilities of the iPod, a state-of-the-art camera, and easy-to-use touchscreen controls. Each person on the planet could now have one of the most powerful technologies most human societies had ever known inside their pockets. Every song, video, news article, and fact on the internet was now—quite literally—at our swiping fingertips.

Jobs's announcement boasted that the iPhone would reinvent the phone, but just over a decade later, it's clear that his invention changed much more. We don't just use our smartphones to call a friend. Nowadays our phones are our photo albums, banks, cameras, newspapers, social clubs, camcorders, weather reports, alarm clocks, marketing departments, music libraries, televisions, movie theaters, travel agents, calendars, flashlights, video conferencing services, inboxes, magazines, maps, traffic reports, calculators, stock market experts, video game consoles, shopping carts, grocery stores, take-out counters, taxi services, step trackers, workout buddies, and porn collections. The smartphones of today have become the most captivating stimulus the human species has ever known. But what effect are these unprecedentedly huge bundles of information, entertainment, and services inside our pockets having on our real-life interactions?

A surprising hint comes from a rather unlikely economic indicator—chewing gum sales. In the decade since the iPhone hit our pockets in 2007, global gum sales have declined about 15 percent. To see why that's relevant, consider the last time you waited in line at the grocery store. If you're like most people, you probably spent at least some of that time staring at your smartphone. In fact, one recent study found that 62 percent of people pulled out a device while waiting in line, and more than 80 percent of those people whipped that device out in less than twenty seconds. Just a few years earlier, standing in line felt very different. Back then, we felt bored—so bored, in fact, that we wound up noticing the world around us. We may have flipped through a rack of magazines or even made an impulse buy of one of the colorful packs of gum that were strategically placed to catch our attention. But even though gum packaging is as enticing as ever, these kinds of last-minute purchases happen significantly less often today. That's because we don't even notice the shocking tabloids or the colorful candy racks. Our eyes are glued to something even more attention-grabbing: all the stuff that's possible on our tiny screens.

That's one of the dirty secrets of much of the content that we see on our phones: Like gum-package designers, app manufacturers and social media engineers want to capture our attention. Their profits rely on keeping our eyeballs glued to those screens by any means necessary, whether that's an annoying notification buzz, small pieces of fun information that come to us on a slot-machine-like, variable-reward schedule, or even a social media algorithm designed to provoke deep anxiety.

Our smartphones are attention-grabbing enough to distract us from impulse candy buys at the supermarket, but our attraction to our devices also prevents us from noticing the things around us that we *should* be paying attention to. That's why phones have infiltrated lots of situations in which we really should be paying attention to something more important. For instance: 80 percent of Americans say they used a cell phone during their last in-person social activity; 36 percent admit to checking their phones while driving in the previous month; 92 percent of college students text during classes; and 55.6 percent of medical staff admit to using their phone in the middle of a cardiopulmonary bypass surgery.

What's even more disturbing is that our cell phones continue to affect us even during those rare times we choose *not* to use them. A growing body of research suggests that there may be a cognitive cost to simply having our attention-grabbing phones nearby. Merely hearing a cell phone notification during a pleasant activity like playing a video game or getting a massage can reduce our enjoyment of that activity by up to 10 percent, even when you don't look at the notification. And research shows that the mere presence of phones—even if they're shut off—can affect cognitive performance. Business school professor Adrian Ward and his colleagues have found that simply having your phone powered off on a nearby table can significantly reduce performance on a cognitively demanding task.

The most ironic effect of our constant connection to our phones is what those devices do to our real-life connections. Jobs thought the iPhone would revolutionize our ability to connect with the people we care about, but the exciting information these devices provide often place an opportunity cost on our IRL (in real life) social connections. One study found that 70 percent of romantic relationships are negatively affected by *phubbing*—a term coined a few years after the introduction of the iPhone—which is the habit of snubbing a person you're hanging out with to check your phone. But a growing number of studies show that phone use hurts the *phubber* as much as the *phubee.* Psychologist Elizabeth Dunn and her colleagues have a number of new studies documenting how much a quick phone peek can affect the enjoyment and connection we feel with the people around us. Dunn and her colleagues have found that using your cell phone during a shared meal can significantly reduce your enjoyment of that meal. Cell phone use during social events also makes us more bored, more distracted, and less connected to the people we're with. Dunn and colleagues have also observed the same effects during family time; parents who used smartphones during an activity with their children reported feeling significantly less connected to their own children. Our phones also prevent us from making connections with new people. In one study, Dunn allowed strangers the opportunity to get to know one another in a waiting room, either with phones present or absent. Shockingly, she found that people smile at one another 30 percent less often when they're around their phones.

These studies suggest that, rather than connecting us, smartphones may be altering many subtle aspects of our natural social fabric—from the dinners we spend connecting with our partner to the joy we experience hanging out with our children to the pleasantries we make with strangers on the street. Perhaps it's no surprise that despite having a device that can connect us to anyone in world, we're all feeling lonelier than ever since these devices have gone mainstream. Indeed, the generation most likely to reap the benefits of this new technology—teens and college students—now face the biggest mental health crisis human society has ever observed, with teens reporting double the rates of depression since 2007. The psychologist and author Jean Twenge has argued for a direct causal link between smartphones and the double-digit increases we've seen in the number of teens reporting that they feel lonely or left out since the iPhone released over a decade ago.

Smartphones are designed to be more than just a phone, but they've evolved to become the most rewarding and entertaining stimulus that human beings have ever known. Jobs was prescient that his new invention would change everything. The original iPhone's descendants are now so captivating that they steal our attention from the things in life that matter most—our partners, our kids, our friends, and all the simple stuff that makes the present moment enjoyable. Psychologists are learning that each of these tiny missed moments of connection and enjoyment are adding up to a growing crisis of distraction, loneliness, and disconnection. Understanding the subtle ways that our screens distract us from the important things in life is the first step to taking our attention back and becoming more intentional about what is worthy of our focused time, one of the most precious resources we have. Just as we've developed rules about the use of cell phones on the road and in the classroom, so, too, can we design better norms for using technology around one another. In doing so, we can begin to rebuild the social fabric that a decade of attention to our devices has subtly eroded.

We're All Connected but No One's in Charge

THOMAS FRIEDMAN

Thomas Friedman is an American political commentator and author. He is a three-time Pulitzer Prize winner who is a weekly columnist for the New York Times.

Welcome to the second inning of one of the world's great technological leaps, the implications of which we're just beginning to understand.

But first, let's acknowledge one thing: The first inning was *amazing.*

It was an inning full of promise, discovery, and marvel. In the early 2000s, a set of technologies came together into platforms, social networks, and software that made connectivity and solving complex problems *fast, virtually free, easy, ubiquitous, and invisible.* Suddenly, more individuals could compete, connect, collaborate, and create with more people, in more ways, from more places, for less money, and with greater ease than ever before. And we sure did!

We became our own filmmakers and reporters; we launched political and social revolutions from our living rooms; we connected with long-lost family and friends; we found the answers to old and new questions with one click; we searched for everything from spouses to news to directions to kindred spirits with our phones; we exposed dictators and branded ourselves. With one touch, we could suddenly call a taxi, direct a taxi, rate a taxi, and pay a taxi—or rent an igloo, rate an igloo, and pay for an igloo in Alaska.

Then, just as suddenly, we found ourselves in the second inning. A critical mass of our interactions had moved to a realm *where we're all connected but no one's in charge.* There are no police officers walking the beat, no courts, no judges, no God who smites evil and rewards good, and certainly no "1-800-Call-If-Putin-Hacks-Your-Election." If someone slimes you on Twitter or Facebook, well, unless it is a death threat, good luck getting it removed, especially if it is done anonymously.

The cool Uber self-driving car killed a pedestrian; the cool Facebook platform enabled Russian troll farms to divide us and inject fake news into our public life; and uncool totalitarian governments learned how to use the same facial recognition tools that can ease your way through passport control to single you out in a crowd for arrest.

And Mark Zuckerberg, who promised to connect us all—and that it would all be good—found himself on the cover of *Wired* magazine, with his face cut, bruised, and bandaged, as if he'd been hit by a fastball. He wasn't alone. In inning two, we started to feel beat up by the same platforms and technologies that had enriched, empowered, and connected our lives.

Silicon Valley, we have a problem.

What to do? For problems like this, I like to consult my teacher and friend Dov Seidman, CEO of LRN, which helps companies and leaders build ethical cultures, and the author of the book *How: Why How We Do Anything Means Everything.*

"The first inning's prevailing ethos was that any technology that makes the world more open by connecting us or makes us more equal by empowering us individually *must,* in and of itself, be a force for good," Seidman began. "But, in inning two, we are coming to grips with the reality that the power to make the world more open and equal is not in the technologies themselves. It all depends on how the tools are designed and how we choose to use them. The same amazing tech that enables people to forge deeper relationships, foster closer communities, and give everyone a voice can also breed isolation, embolden racists, and empower digital bullies and nefarious actors."

Equally important, Seidman added, these "unprecedented and valuable tools of connection" are being used with great accuracy and potency "to assault the foundations of what makes our democracies vibrant, capitalism dynamic, and our societies healthy—namely, truth and trust."

They have also been used "to assault our personal foundations—our privacy and sense of identity," Seidman said. "It is one thing to use our data to enable better shopping experiences, but when my beliefs and attitudes are mined and manipulated for someone's political campaign, a campaign that may be antithetical to my beliefs, that is deeply harmful and unmooring." So what to do? "Precisely because we are in just the beginning of a technological revolution with a long, uncertain, up-and-down road ahead, we need to start by *pausing* to reflect on how our world, reshaped by these technologies, operates differently—and on the kind of values and leadership we will need to realize their promise."

Values are more vital now than ever, Seidman insisted, "because sustainable values are what anchor us in a storm, and because values propel and guide us when our lives are profoundly disrupted. They help us make the hard decisions." Hard decisions abound because everything is now connected. "The world is fused. So there is no place anymore to stand to the side and claim neutrality—to say, 'I am just a businessperson' or 'I am just running a platform.'"

No way. "Once you see that your technologies are having unintended consequences, you cannot maintain your neutrality—especially when you've become so central to the lives of billions of people."

In the fused world, Seidman said, "the business of business is no longer just business. The business of business is now society. And, therefore, how you take or don't take responsibility for what your technology enables or for what happens on your platforms is inescapable. This is the emerging expectation of users—*real people*—who've entrusted so much of their inner lives to these powerful companies."

To be sure, Facebook, Twitter, and YouTube should all be commended for trying to find engineering solutions to prevent them from being hacked and weaponized. "But this is not just an engineering problem or just a business-model problem," he said. "Software solutions can increase our confidence that we can stay a step ahead of the bad guys. But, fundamentally, it will take more 'moralware' to regain our trust. Only one kind of leadership can respond to this kind of problem—moral leadership."

What does moral leadership look like here?

"Moral leadership means truly putting people first and making whatever sacrifices that entails," said Seidman. "That means not always competing on shallow things and quantity—on how much time people spend on your platform—but on quality and depth. It means seeing and treating people not just as 'users' or 'clicks' but as 'citizens' who are worthy of being accurately informed to make their best choices. It means not just trying to shift people from one click to another, from one video to another, but instead trying to elevate them in ways that deepen our connections and enrich our conversations."

It means, Seidman continued, being "fully transparent about how you operate and make decisions that affect them—all the ways in which you're monetizing their data. It means having the courage to publish explicit standards of quality and expectations of conduct, and fighting to maintain them however inconvenient. It means having the humility to ask for help even from your critics. It means promoting civility and decency, making the opposite unwelcome. It means being truly bold—proclaiming, for example, that you will not sleep until you're certain that our next democratic election won't be hacked."

At the height of the Cold War, when the world was threatened by spreading Communism and rising walls, President John F. Kennedy vowed to "pay any price and bear any burden" to ensure the success of liberty. Today, falling walls and spreading webs—which criminals and nations can use to poison democratic societies—are becoming the biggest threat to the success of liberty. You will know that the good guys are winning when you see big tech companies rise to Kennedy's challenge—to pay any price and bear any burden to protect us from the downsides of the technologies they've created.

Just once I'd like to see Zuckerberg look into a camera and say: "I will take Facebook stock down to $1 if that is what it takes to ensure that we're never again an engine for the perversion of democracy in any country, starting with my own. Facebook is not going to accept any more political ads until we have the resources to fact-check them all."

I doubt he'll do that, though, because his priorities are profits and power, and he seems quite ready to hurt American democracy to get them.

Welcome to the second inning.

From the New York Times.

Technology Can Augment Our Humanity or Consume It

ARIANNA HUFFINGTON

Arianna Huffington is the founder of the Huffington Post, *the founder and CEO of Thrive Global, and the author of fifteen books, including, most recently,* Thrive *and* The Sleep Revolution.

Since 2016, our relationship with technology has experienced a profound shift. Everything about that relationship is being called into question. But even as our attitudes toward technology have been rapidly changing, so has technology. We're now at an inflection point in which we have to decide what we want out of it and what we want its role to be in our society and our individual lives.

The Awakening

Our clearest wake-up call about technology was the 2016 presidential election. Conversations about technology's effects have been happening for years, but the revelations about how its use helped undermine the election pushed those conversations into the collective consciousness. By the time the Mueller report was released in April 2019, it was clear, as its conclusion said, that "the Russian government interfered in the 2016 presidential election in sweeping and systematic fashion."

Until that point, the narrative about technology and democracy had been about the power of social media to connect people, coalesce popular sentiment, and challenge governments and traditional power structures. That crystallized in 2010, when social media was credited with a crucial role in driving the Arab Spring.

A lot has changed since then. In addition to exposing the truth, social media can also undermine it. "The possibility of creating an alternative narrative is one people didn't consider," said historian Anne Applebaum, "and it turns out people in authoritarian regimes are quite good at it." Governments went from simply trying to turn off the internet, as former Egyptian President Hosni Mubarak did during the Tahrir Square uprising, to using its power themselves. "In seven years," wrote professor and technology security expert Zeynep Tufekci, "digital technologies have gone from being hailed as tools of freedom and change to being blamed for upheavals in Western democracies—for enabling increased polarization, rising authoritarianism, and meddling in national elections by Russia and others."

By 2019, the change in attitude was clear. According to a survey by the Pew Research Center, from 2015 to 2019, the share of Americans who believed that technology was positively impacting the country fell from 71 percent to 50 percent, and the share of those who saw technology's impact as negative nearly doubled, rising from 17 percent to 33 percent.

The Reckoning

While we grappled with social media's impact on how we govern ourselves collectively, we also grew more aware of how our phones and screens affect how we govern ourselves individually. The tenth anniversary of the iPhone was in 2017, and by the start of the year, over three-quarters of Americans owned smartphones. But unquestioning enthusiasm was giving way to growing skepticism—not just among users but among tech executives, whose rhetoric shifted from triumphal to confessional.

Meanwhile, users were becoming increasingly stressed. A 2018 Gallup poll found that 67 percent of American employees felt burned out at work. The next year, the World Health Organization added "burnout" to its International Classification of Diseases and Related Health Problems. The pace of our lives was accelerating beyond our capacity to keep up. People began to wonder why they always felt harried and behind, perpetually trapped in what researchers call "time famine." More and more people pointed the finger at the very devices they had thought were saving them so much time. It was like the dramatic scene in horror movies where the cop tells the frantic family, "We've traced the call—and it's coming from inside the house!"

The last several years have seen an avalanche of science showing how profoundly our relationship with technology affects our mental and physical well-being. Research shows associations between screen use and higher rates of stress, anxiety, depression, and even suicide. A 2019 study by the Cincinnati Children's Hospital Medical Center found that screen use in young children actually changed their brains, creating differences in the regions that govern language and self-regulation. A 2019 survey by Common Sense Media found that 45 percent of parents feel addicted to their devices, and 39 percent of kids wish their parents would get off them. Our use of technology has become a serious public health issue.

Technology has connected everybody to everybody. At the same time, we've found ourselves in what former Surgeon General Vivek Murthy calls a "loneliness epidemic." That drew headlines when Murthy said it in 2017, but it's no longer disputed. A 2019 survey by Cigna found that more than half the country reported feeling lonely. And the loneliest generations? The ones that had been most surrounded by technology from birth: Gen Zers and millennials. We've also learned how dangerous loneliness is: A study by researchers from Brigham Young University found that it comes with the same mortality risk as smoking fifteen cigarettes a day.

We're hardwired to connect, and we've wired our whole world so that nobody is ever out of reach. We've filled in all the space and connected all the interstices—and yet we're all trapped in our own silos, disconnected from one another and ourselves. Instead of fulfilling our fundamental need for connection and meaning—which we only get from real connection with one another—we're all, as MIT professor Sherry Turkle puts it, "alone together."

Recalibrating

That brings us to our current moment. When you look back, it's hard to believe just how different it is from where we were only a few years ago. The conversation about technology is now dominated by how to manage it, how to control it, and how to deal with its constant presence in our lives.

Technology allows us to do amazing things and has led to dramatic strides in alleviating human suffering around the world. It will always be with us. Cyberspace is now just space. But it's up to us to decide what we want out of technology.

As the dust settles from our first decade of being connected 24/7, the picture is becoming clearer. We want technology that works with us, for our best interests and well-being, and not against us. We want human-centered technology that augments our humanity instead of consuming it. We want technology that enhances our attention instead of mining it. We might accept driverless cars, but we don't want driverless humans. We want to be back in the driver's seat ourselves.

A 2019 poll found that over 52 percent of Americans intentionally look for ways to unplug from technology. And those who are most adept and comfortable with technology are the most desperate to find ways to escape from it—especially members of Generation Z, who are even more likely than millennials to be always reachable.

It turns out that one of the ways we can take control of our relationship with technology is through more technology. Our psychological levers are being pushed and pulled every day. Mostly they're used to induce want, FOMO (fear of missing out), and the desire to spend. But those same levers can be used to help us adopt and nurture healthier habits. The same technology that is used to hook us to social media and gaming can be used instead to unhook us and help us rebuild the boundaries of humanity. Although we hear about artificial intelligence (AI) mostly in the context of large-scale disruption of the workplace, its most exciting use, to me, is at the individual level. Using data-driven feedback loops, AI can help us create new and healthier habits. That technology exists now, and more of it is being developed. According to a 2019 report by Global Market Insights, the global digital health market will hit $500 billion by 2025. This will be about more than just tracking. If we want to take control of technology, and our well-being, we can.

As we redefine our individual relationship to technology, we can rebalance our collective relationship with it. We need to secure our elections from hacking. We need stricter regulations for political ads and laws safeguarding our data and our privacy. But for our collective to have true integrity, we need to first create a healthy relationship with technology in our own lives. Then we'll be able to maximize our wisdom to deal with our collective challenges.

Shock Therapy

TIM WU

Tim Wu is a professor at Columbia Law School and a contributing opinion writer for the New York Times. *He formerly worked at the White House and is a member of the American Academy of Arts and Sciences.*

I want to take a very broad perspective on what are called "technological shocks"—indeed, the centuries-long view. We are a technophilic society, one that loves and adores technological progress, sometimes irrationally. But the truth is that really big inventions, like the printing press, the steam engine, the machine gun, and the internet have both upsides and downsides. They make new things possible, but they also tend to undo settled expectations and create chaos. On the other hand, shutting down technological progress has other severe downsides. The real question is not *will* there be major technological changes, but whether societies can learn to better handle technological shocks.

There's a lot to learn because, when we take a look at the history, most of the time we have blown it. Consider how things stood a century ago, in the early to mid-twentieth century: We sometimes think that technology is changing faster today than ever before, but that is a good example of the self-delusion every age indulges in. The 1890s through the 1920s witnessed the invention of airplanes, home electricity, radio broadcasting, engines, bombs, tanks, and the machine gun. Take that, Instagram and Airbnb.

It is no surprise that the invention of so much technology in such a short period was completely destabilizing in economic, social, and military terms. And it is extraordinary, in retrospect, to see how badly it all went. That period, and the period immediately afterward, witnessed terrible labor violence, two depressions, the rise of totalitarianism, two world wars, several genocides, and other mass killings of unprecedented volume. If these horrors were not exactly caused by the wondrous new technologies of the age, they were certainly aided and abetted by them. That's what might be called the ultimate failure to adapt.

Colonial history provides another example of a different kind of technological shock. The Western powers mastered the use of new transportation and military technologies—gunpowder and ships, most prominently. With those technologies, they were suddenly able to conquer and subjugate entire continents and enslave whole populations, a process so traumatic that, centuries later, the scars remain.

The inventions just described were like the adding of a catalyst, upending settled equilibria—creating what chemists call rapid reactions and what most people call explosions. The military inventions upended whatever deterrence equilibrium existed, giving some groups (like the Spanish, Portuguese, Germans, and Japanese) good reason to think they might decimate others. Similarly, a major economic change, like rapid-fire industrialization, let a new set of companies wipe out the old ones and allowed employers to exploit their employees in new ways. While an industrial massacre is different than an actual massacre, both yield real suffering.

If we know the past effects of technology shocks, why aren't we more careful? After all, we know that rivers flood and volcanos erupt, and we at least take some efforts to mitigate the risks. Why not try to do more to limit the social effects of big inventions?

Well, it turns out that some civilizations have, in fact, been extremely cautious about the potential damage caused by technological shocks. Consider the Ming dynasty of China (1368–1644). The Ming arose after the suffering and chaos of the Mongolian invasions, which was its own kind of global technological shock. If Hong Wu, the first Ming emperor, had a campaign slogan, it would have been "a return to normalcy." He and his successors tried hard to prevent anything from changing and sought to mimic historic golden periods. If you called a Ming emperor "backward looking," he would have been pleased with the compliment. The Ming did so, in part, based on the fear that technological change would create unrest and suffering.

At the risk of stating the obvious, technology-repressive civilizations like the Ming dynasty or medieval Europe, at their worst, have downsides. Some have described the Ming dynasty as the world's first totalitarian state. Suppressing technological change tends to require a level of state control over life that is not pleasant and creates the risk of another kind of technological shock—facing a conquering army equipped with much more advanced weapons. In China's case, suppression of technological development led to its partial colonization by Western powers, untold suffering in the twentieth century, and its invasion by the Japanese.

We don't have the problem of Ming China. Instead, today's dominant civilizations have the opposite orientation—extreme technophilia. Western civilization, especially America, is exceptionally forward looking. We are always imagining utopian futures, believing that "the best is yet to come." The phrase "scientific progress" has an almost talismanic allure to it, and to say someone is "backward looking" is usually insulting. Even Christian conservatives, who may revere the past, also look forward to the glories of the coming Kingdom of Christ and, in fact, have often been among the first to master new media technologies. Indeed, it may be that our entire forward-looking perspective has its roots in a religious longing to transcend this material world.

The result is that technophilia can be quite extreme; as social critic Neil Postman put it in 1992, we "gaze on technology as a lover does on his beloved, seeing it without blemish and entertaining no apprehension for the future." If that is not exactly us right now, it was us circa 1992. We seem unable to imagine what might go wrong—Facebook, after all, was just about making friends, right? Postman, in his book *Technopoly*, opens with the startling observation (credited to Egyptian King Thamus) that even the invention of *writing* had costs as well as benefits. According to Plato, Thamus said, "What you have discovered is a receipt for recollection, not for memory. And as for wisdom, your pupils will have the reputation for it without the reality. . . . And because they are filled with the conceit of wisdom instead of real wisdom, they will be a burden to society." (That sounds a bit like Google searching, actually. You can seem to know a lot, when you really know very little at all.)

As I say, not all civilizations are this way. China, today, is arguably even more technophilic than the United States. Perhaps arising out of its long humiliation in the nineteenth and twentieth centuries, its total embrace of technologies can make even American fanboys look hesitant. This is a country that has declared that it will be the world's master of artificial intelligence by the year 2030 and is embarking on all kinds of dubious projects to achieve that—including mandating that courts use robots to decide civil and criminal cases.

So if the alternatives are extreme technophobia (Ming China) or extreme technophilia (ourselves until recently), I'd like to believe we can do better—by creating a resilient civilization capable of surviving periods of very rapid technological change without tipping into class warfare, severe economic depression, violent revolution, mass dislocation, colonization, or catastrophic warfare.

Societies can become resilient in two main ways. First, they can take steps to buffer and mitigate the effects of social dislocation. For example, since technological shocks have historically led to a major consolidation of wealth and the rise of new monopoly powers, societies can dissipate the shock by breaking the monopolies and ensuring there are measures for redistributing wealth. Second, instead of being uniformly technophilic and -phobic, it might be wise for societies to go through cycles. One cycle of inventing a whole bunch of new stuff could be followed by another cycle of fixing all the damage that's been done, then repeat.

Judging by historic standards, we could be doing worse. Thanks to things like social safety nets, we are probably a little more resilient now than in the twentieth century. But we're not out of the woods yet, and things could still get much worse before they get better. First, the rise of the tech platforms has already triggered forms of struggle and borderline class warfare (as in Doug Rushkoff's book *Throwing Rocks at the Google Bus*) as tech disrupts industries at the periphery of the economy, including advertising and some retail. If larger, more significant industries began to fall—say, for example, the entire automobile industry—the economic uncertainty could be nothing short of explosive.

Second, the invading-army kind of technological shock is not a joke. I don't believe that we should prop up Facebook and Google in the hope that their monopoly power is some kind of defense against the Chinese or Russian menace. But I do think that the country needs to be able to deter and better prevent things like foreign manipulation of its elections and the stirring of domestic unrest.

I do see signs that we have moved to a less technophilic period and are taking a hard look at where we've gone wrong, what damage we've done, and what we might fix. Now we might be ready to repair the damage. We are in the fix-it cycle, and I hope a generation of engineers, scientists, and politicians will lead us out of technological shock, back to stability.

Tech and Creative Destruction

JONATHAN TAPLIN

Jonathan Taplin is director emeritus of the Annenberg Innovation Lab at the University of Southern California, and he is the author of Move Fast and Break Things: How Facebook, Google, and Amazon Cornered Culture and Undermined Democracy.

Reform movements demonstrate that history is made by abrupt transitions. The 1890s are remembered as the Gilded Age, when plutocrats like J. P. Morgan and John D. Rockefeller asserted control over the US economy and politics. By 1906, both Rockefeller and Morgan were being forced by antitrust regulators to break up their vast holdings. When I first published *Move Fast and Break Things* in April 2017, I thought we were in another plutocratic era. Now, just three years later, as I write this, we are in fact at the beginning of a profound change in how we view tech monopolies.

Since 2017, the European Union has fined Google $9.3 billion for abusing its monopoly power, and there is increasing evidence that American politicians and regulators are also open to new regulation of Google, Facebook, and Amazon. We have realized that we have entrusted them with much of our most intimate data, and they may not be worthy of that trust. We use social networks to share personal news and photographs of ourselves and loved ones. Meanwhile, the National Center for Missing and Exploited Children reports that the number of child sexual abuse images online exploded from three thousand in 1998 to forty-five million in 2018. The center also found that one platform, Facebook Messenger, carried over two-thirds of those disturbing images.

Use of the internet as a propaganda machine instead of simply a benign, ever-flowing source of information has changed the political communications game. The hijacking of social media as a propaganda organ is distinctly different from partisan radio and television. To begin with, our smartphones are with us every waking hour, whereas TV and radio are not regularly consumed in the workplace. Every day, we check our phones 150 times, and Facebook alone gets fifty-four minutes of our time on average. The ability of notifications to interrupt other activities, combined with the ability to deliver random rewards like a slot machine, leads to addictive behaviors that make us perfect receptors and transmitters of propaganda. What Trump's former campaign manager Stephen Bannon understood was that when you link a fake news article ("The Pope Endorses Trump") to a Facebook page and then promote it with a million bots, you can move the article to the top of Google search and Facebook's "What's Trending" almost instantly. As former Google engineer Tristan Harris has shown, "Bot networks are used to intimidate users, fabricate social consensus, manipulate #trending topics, propagate disinformation, and manipulate public opinion." They are very effective. As *BuzzFeed* reports, "In the final three months of the [2016] US presidential campaign, the top-performing fake election news stories on Facebook generated more engagement than the top stories from major news outlets such as the *New York Times, Washington Post, Huffington Post,* NBC News, and others."

As I write this, neither Facebook nor Google has fully disclosed the extent of foreign manipulation of our last two elections via their platforms. Facebook has made some early gestures to make their ad platform more transparent, but as the *New York Times* recently reported, "The public may have little more insight into disinformation campaigns on the social network heading into the 2020 presidential election than they had in 2016." We know that Google's AdSense software, which provided much of the revenue to the Eastern European teams that were flooding the web with fake news, knew both the IP address and, in many cases, the bank account information of the fake news providers.

The last ten years have seen the wholesale destruction of the creative economy—journalists, musicians, authors, and filmmakers—wrought by three tech monopolies: Google, Facebook, and Amazon. Their dominance in artificial intelligence will extend this "creative destruction" to much of the service economy, including transportation, medicine, and retail. There is not a single politician in America talking about this, and when the flood of unemployment brought about by the artificial intelligence revolution is upon us, we will not be ready. Treasury Secretary Steven Mnuchin was quoted as saying that the robotics and AI revolution would not arrive for a hundred years: "In terms of artificial intelligence taking over American jobs—I think we're, like, so far away from that that it is not even on my radar screen." But Mnuchin's former employer, Goldman Sachs, reported that self-driving vehicles could eliminate three hundred thousand jobs per year starting in 2022. Both sides of this argument cannot be true, but we are forging ahead with a vision of an AI universe with almost no political debate. We know this is true because of the deafening silence from the politicians in the last ten years, as 50 percent of the jobs in journalism were eliminated and revenues at both music companies and newspapers fell by 70 percent. Who was there to speak for the creative workers of the world?

The companies that will win the AI race are already in the forefront: Google, Facebook, and Amazon. As AI venture capitalist Kai-Fu Lee wrote, "AI is an industry in which strength begets strength: The more data you have, the better your product; the better your product, the more data you can collect; the more data you can collect, the more talent you can attract; the more talent you can attract, the better your product." Google, Facebook, and Amazon are already pushing out of tech into other sectors of the economy, as Amazon's acquisition of Whole Foods demonstrates. Google's life sciences division, Verily, is producing glucose-monitoring contact lenses for diabetics, wrist computers that read diagnostic nanoparticles injected in the bloodstream, implantable devices that modify electrical signals that pass along nerves, medication robots, human augmentation, and human brain simulation devices. Google's autonomous car division is already working with Avis to manage their forthcoming self-driving car fleet. As for Facebook's brand extension plans into video, they bid $800 million for the worldwide rights to broadcast Indian cricket on their platform, only to be outbid by Rupert Murdoch's Star India. These are just the start of many initiatives to extend the tech giant's technologies into many parts of the American economy.

I think big changes could happen if we approach the problem of monopolization of the internet with honesty, a sense of history, and a determination to protect what we all agree is important: our cultural inheritance. We all need the access to information the internet provides, but we need to be able to share information about ourselves with our friends without unwittingly supporting a corporation's profits. Facebook and Google must be willing to alter their business model to protect our privacy and help thousands of artists create a sustainable culture for the centuries, not just make a few software designers billionaires. We need to amend the Section 230 "safe harbor" provision of the Telecommunications Act of 1996 to provide for a "take down, stay down" provision. If musicians do not want their work on YouTube or Facebook for free, they should be able to file a takedown notice, and then it becomes the responsibility of the platform to block that content from ever being uploaded. All the tools needed to make this happen already exist.

But we also must understand that the people who run Google, Facebook, and Amazon are just at the beginning of a long project to change our world, so this battle is just beginning. Yuval Noah Harari calls their project "Dataism":

> *Dataists further believe that given enough biometric data and computing power, this all-encompassing system could understand humans much better than we understand ourselves. Once that happens, humans will lose their authority, and humanist practices such as democratic elections will become as obsolete as rain dances and flint knives.*

We need to confront this techno-determinism with real solutions before it is too late. I undertake the pursuit of these solutions with both optimism and humility. Optimism because I believe in the power of rock and roll, books, and movies to upset the world. As the late Toni Morrison observed, "The history of art, whether it's in music or written or what have you, has always been bloody, because dictators and people in office and people who want to control and deceive know *exactly* the people who will disturb their plans. And those people are artists. They're the ones that sing the truth. And that is something that society has got to protect." I know that brave and passionate art is worth protecting and is more than just clickbait for global advertising monopolies. I know it can change lives.

I hope to see Tim Berners-Lee's dream of a "re-decentralized" internet, one that's much less dependent on surveillance marketing and that allows creative artists to take advantage of the zero-marginal-cost economics of the web by forming nonprofit distribution cooperatives. I have no illusion that the existing business structures of cultural marketing will go away, but my hope is that we can build a parallel structure that will benefit all creators.

The only way this will happen is if, in Peter Thiel's "deadly race between politics and technology," the people's voice (politics) will win. Google, Amazon, and Facebook may seem like benevolent plutocrats, but the time for plutocracy is over.

Mad Men and Math Men

KEN AULETTA

Ken Auletta is a bestselling author. Since 1992, he has written the Annals of Communications for the New Yorker.

There are basically two ways of reacting to digital disruption. There are those who lean back, feeling helpless, paralyzed, moaning about how awful and unfair it is that they have been victimized by uncaring forces beyond their control. Then there are those who lean forward, seeing change as an opportunity, not just a problem. So they energetically innovate, constantly trying new approaches to salvage and shape their world.

The wise ones who lean forward don't romanticize either the old world or the new. Leaning forward, they understand that while data and algorithms and artificial intelligence are of great value, they lack the instinct, creativity, and wisdom of humans. They are aware that Facebook's algorithms failed to block fake news or malicious and viral videos, requiring Facebook to belatedly hire thousands of human "curators" to mount a defense.

The advertising industry has been late to feel the force of digital disruption but is now reeling. We may hate most ads and feel they are noisy interruptions, blatant attempts to manipulate our emotions. But the estimated two trillion or so dollars spent worldwide on advertising and marketing matters. To stay relevant, advertisers believe they must move quickly to rely on the Big Data the digital world generates in order to become more intimate with consumers. To do that, they must breach the privacy wall.

Once, Mad Men ruled advertising. Creative geniuses like Bill Bernbach followed their instincts to create ads. They've now been eclipsed by Math Men—the engineers and data scientists whose province is machines, algorithms, pureed data, and artificial intelligence.

In the advertising world, Big Data is the holy grail. It enables marketers to target messages to individuals rather than general groups, creating what's called "addressable advertising." Only the digital giants possess state-of-the-art Big Data. Google and Facebook each has a market value exceeding the combined value of the six largest advertising and marketing holding companies. Together, they claim six out of every ten dollars spent on digital advertising, and nine out of ten new digital ad dollars. It took Facebook and Google about five years before they figured out how to generate revenue, and today roughly 95 percent of Facebook's dollars and almost 90 percent of Google's comes from advertising. Facebook alone generates more ad dollars than all of America's newspapers, and Google has twice the ad revenues of Facebook.

"The game is no longer about sending you a mail-order catalogue or even about targeting online advertising," as Shoshana Zuboff, a professor of business administration at the Harvard Business School, wrote on faz.net in 2016. "The game is selling access to the real-time flow of your daily life—your reality—in order to directly influence and modify your behavior for profit." Success at this game flows to those with the "ability to predict the future—specifically the future of behavior." She dubs this "surveillance capitalism."

According to *ProPublica*'s series "Breaking the Black Box," Facebook "has a particularly comprehensive set of dossiers on its more than two billion members. Every time a Facebook member likes a post, tags a photo, updates their favorite movies in their profile, posts a comment about a politician, or changes their relationship status, Facebook logs it. . . . When they use Instagram or WhatsApp on their phone, which are both owned by Facebook, they contribute more data to Facebook's dossier." The company offers advertisers more than thirteen hundred categories for ad targeting.

Google, for its part, has merged all the data it collects from its search, YouTube, and other services. It has introduced an "About Me" page, which offers advertisers your date of birth, phone number, where you work, mailing address, education, where you've traveled, and your nickname, photo, and email address. Amazon knows even more about you. Since it's the world's largest store and sees what you've actually purchased, its data about you are unrivaled. Amazon reaches beyond what interests you (revealed by a Google search) or what your friends are saying (on Facebook) to what you actually purchase. With Amazon's Alexa, it has an agent in your home that not only knows what you bought but when you wake up, what you watch, read, listen to, ask for, and eat. And Amazon is aggressively building up its once-meager ad sales, giving it an incentive to exploit its data.

Engineers and data scientists believe there is nobility in their quest. By offering individualized marketing messages, they are trading something of value in exchange for a consumer's attention. They also start from the principle, as the TV networks did, that advertising allows their product to be "free." Of course, as their audience swells, so does their data. Sandy Parakilas, who was Facebook's operations manager on its platform team from 2011 to 2012, put it this way in a scathing November 2017 op-ed for the

New York Times: "The more data it has on offer, the more value it creates for advertisers. That means it has no incentive to police the collection or use of that data—except when negative press or regulators are involved." For the engineers, the privacy issue—like "fake news" and even fraud—has been largely irrelevant.

But to thrash just Facebook and Google is to miss the larger truth. *Everyone* in advertising strives to eliminate risk by perfecting the targeting of data. Foremost among the concerns of marketers is not protecting privacy but protecting and expanding their business. Like Facebook and Google, ad agencies and their clients aim to massage data to better identify potential customers and influence consumer behavior. Agencies compete to proclaim their own Big Data horde. WPP's GroupM, the largest media agency, has quietly assembled what it calls its "secret sauce," a collection of forty thousand personally identifiable attributes it plans to retain on two hundred million adult Americans. To parade their sensitivity to privacy, agencies reassuringly boast that they don't know the names of people in their data bank. But they do have your IP address, which yields abundant information, including where you live. For marketers, the promise of being able to track online behavior, according to former senior GroupM executive Brian Lesser, is that "we know what you want even before you know you want it."

Keith Weed—who until 2019 oversaw marketing and communications for Unilever, one of the world's largest advertisers—described how mobile phones have elevated data as a marketing tool. "When I started in marketing," he said, "we were using secondhand data which was three months old. Now with the good old mobile, I have individualized data on people. You don't need to know their names. . . . You know their telephone number. You know where they live because it's the same location as their PC." Weed knows what times of the day you usually browse, watch videos, answer email, travel to the office, and what travel routes you take. "From your mobile, I know whether you stay in four-star or two-star hotels, whether you go to train stations or airports. I use these insights along with what you're browsing on your PC. I know whether you're interested in horses or holidays in the Caribbean." By using programmatic computers to buy ads targeting these individuals, he says, Unilever can "create a hundred thousand permutations of the same ad," as they recently did with a thirty-second TV ad for Axe toiletries aimed at young men in Brazil. The more Keith Weed knows about a consumer, the better he can target a sale.

But Math Men's adoration of data—coupled with their truculence and an arrogant conviction that their "science" is nearly flawless—has aroused government anger, much as Microsoft did two decades ago. With a chorus of marketers and citizens and governments roaring their concern, the limitations of Math Men loom large. Suddenly, governments in the United States are almost as alive to privacy dangers as those in Western Europe. They confronted Facebook by asking how the political-data company Cambridge Analytica, employed by Donald Trump's presidential campaign, was able to snatch personal data from eighty-seven million individual Facebook profiles. Was Facebook blind—or deliberately mute? Why, they are really asking, should we believe in the infallibility of your machines and your willingness to protect our privacy?

The credibility of these digital giants was further subverted when Russian trolls proved how easy it was to disseminate "fake news" on social networks. When told that Facebook's mechanized defenses had failed to screen out disinformation planted on the social network to sabotage Hillary Clinton's presidential campaign, Mark Zuckerberg at first publicly dismissed the assertion as "pretty crazy," then later conceded he was wrong.

By the spring of 2018, Facebook had lost control of its narrative. Its original declared mission—to "connect people" and "build a global community"—had been replaced by an implicit new narrative: *We connect advertisers to people.* The humiliating furor this provoked has not subverted the faith among Math Men that their "science" will prevail. They believe advertising will be further transformed by new scientific advances like artificial intelligence that will allow machines to customize ads, marginalizing human creativity. With algorithms creating profiles of individuals, Airbnb's then-chief marketing officer, Jonathan Mildenhall, told me, "Brands can engineer without the need for human creativity." Machines will craft ads, just as machines will drive cars. But the ad community is increasingly mistrustful of the machines and of Facebook and Google. During a presentation at Advertising Week in New York in September 2017, Keith Weed offered a report to Facebook and Google. He gave them a mere C for policing ad fraud, and a harsher F for cross-platform transparency, insisting, "We've got to see over the walled gardens."

Ad agencies and advertisers have long been uneasy not just with the "walled gardens" of Facebook and Google but with their unwillingness to allow an independent company to monitor their results, as Nielsen does for TV and Comscore does online. This mistrust escalated starting in 2016, when it emerged that Facebook and Google charged advertisers for ads that tricked other machines to believe an ad message was seen by humans when it was not. Advertiser confidence in Facebook was further jolted later in 2016, when it was revealed that the Math Men at Facebook overestimated the average time viewers spent watching video by up to 80 percent. And in 2017, Math Men took another beating when news broke that Google's YouTube and Facebook's machines were inserting friendly ads on unfriendly platforms, including racist and porn sites. These were ads targeted by keywords like "Confederacy" or "race"; placing an ad on a history site might locate it on a Nazi-history site.

Now that mistrust has gone viral. A powerful case for more government regulation of the digital giants was made by the *Economist*, a classically conservative publication that, in 1999, also endorsed the government's antitrust prosecution of Microsoft. The magazine editorialized, in May 2017, that governments must better police the five digital giants—Facebook, Google, Amazon, Apple, and Microsoft—because "old ways of thinking about competition, devised in the era of oil, look outdated in what has come to be called the 'data economy.'" Inevitably, an abundance of data alters the nature of competition, allowing companies to benefit from network effects, with users multiplying and companies amassing wealth to swallow potential competitors.

A Who's Who of Silicon Valley notables—Tim Berners-Lee, Tim Cook, Ev Williams, Sean Parker, and Tony Fadell, among others—have also harshly criticized the social harm imposed by the digital giants. Marc Benioff, the CEO of Salesforce—echoing similar sentiments expressed by Berners-Lee—has said, "The government is going to have to be involved. You do it exactly the same way you regulated the cigarette industry."

Cries for regulating the digital giants are almost as loud today as they were to break up Microsoft in the late nineties. Congress insisted that Facebook's Zuckerberg, not his minions, testify. The Federal Trade Commission is investigating Facebook's manipulation of user data. Thirty-seven state attorneys general joined a demand to learn how Facebook safeguards privacy. The European Union has imposed huge fines on Google and wants to inspect Google's crown jewels—its search algorithms—claiming that its search results are skewed to favor their own sites. The European Union's twenty-eight countries in May 2018 imposed a General Data Protection Regulation to protect the privacy of users, requiring that citizens must choose to *opt in* before companies can horde their data. Here's where advertisers and the digital giants lock arms: They speak with one voice in opposing *opt-in* legislation, which would deny access to data without the permission of users. Consumers who wish to deny advertisers access to their data must voluntarily *opt out* through a cumbersome, confusing series of online steps.

Advertising matters because without its subsidies most newspapers and magazines and most of television—not to mention Facebook and Google and most digital media—would perish. But those who rely on the advertising business have reasons to remain tone-deaf to privacy concerns. Remember these twinned, rarely acknowledged truisms: More data probably equals less privacy, while more privacy equals less advertising revenue.

If Big Data's use is circumscribed to protect privacy, the advertising business will suffer. And the disruption that earlier slammed the music, newspaper, magazine, taxi, and retail industries is also upending advertising. Agencies are being challenged by a host of competitive frenemies: consulting and public-relations companies that have jumped into their business; platform customers like Google and Facebook as well as the *Times*, NBC, and *BuzzFeed*, that now double as ad agencies by talking directly to their clients; and business clients that increasingly perform advertising functions in-house.

The foremost frenemy is the annoyed public. Citizens protest aggravating, interruptive advertising, particularly on their personal mobile devices. An estimated 20 percent of Americans, and one-third of Western Europeans, employ ad-blocker software. More than half of those who record programs on their DVRs choose to skip the ads. Netflix and Amazon, among others, have accustomed viewers to watch what they want when they want, without commercial interruption—posing an existential threat not just to agencies but to Facebook and the ad revenues on which most media continue to rely.

Understandably, those dependent on ad dollars quake. In this core concern, at least, Mad Men and Math Men are alike. Not "enemies of the people," as Donald Trump, who is undoubtedly unaware of Ibsen's haunting play, likes to say of the press. But worthy of our fierce attention.

Adapted from Frenemies: The Epic Disruption of the Ad Business (and Everything Else) *(New York: Penguin Press, 2018).*

The Three Sacreds—and Their Disruptions

HOWARD GARDNER

Howard Gardner is the Hobbs Research Professor of Cognition and Education at the Harvard Graduate School of Education. Senior director of Harvard Project Zero and cofounder of the Good Project, he has studied and written extensively about intelligence, creativity, leadership, and professional ethics.

Most human beings have things that they consider to be very special—worth preserving, even worth fighting for. For much of human history and for many today, these special items include members of one's family, one's religion, one's nation, and certain tangible objects, from photographs to heirlooms. At the same time, all too often, we take these things for granted. Indeed, we may only recognize their importance when they are in jeopardy—when we fear that we may lose them, or indeed, when we no longer possess them.

While I value each of these treasures, of late I have come to realize that there are other human inventions that I consider sacred—the professions, institutions of higher education, and the pursuit of knowledge in a disinterested fashion. They, too, have been disrupted, and if we do not act, we are at risk of losing them.

Each of these three disruptions was magnified—perhaps multiplied multifold—by high-speed communication, available to everyone. Anyone can say whatever he or she likes, without consequence, and powerful computational algorithms and devices accumulate as much data as possible about all human beings, for the purposes of consumerism and control, far too often at the expense of privacy and accuracy.

In recent centuries, individuals deemed professionals have been charged with handling human needs—health; justice; verifiable knowledge of how the physical, natural, and social world works; and the attainment of important new knowledge. Professionals like doctors, lawyers, journalists, and professors have been given power and prestige with the understanding that they will behave fairly, serving the wider community instead of filling their own pockets or favoring their own personal causes.

Over the course of the most recent millennium, institutions of higher learning have also emerged, with two main purposes—to make sure that humanity's current knowledge and skills are preserved and passed on to the next generation and to equip a group of these individuals to add to the corpus of human knowledge, correcting the present record and opening up new areas of study and discovery.

Equally important is the pursuit of knowledge—not for an ulterior motive, but to learn and understand as much as possible of our world and our entire universe. Crucially, that knowledge must be as accurate as possible; when it is not accurate, errors or misconceptions must be acknowledged and the record corrected. Central to the pursuit of knowledge is the commitment to truthfulness. It's no accident that two of our oldest educational institutions, Harvard and Yale, feature *veritas* on their insignia. When truth is minimized or ridiculed, the pursuit of knowledge is impossible.

The decline of the professions is due in part to the self-centeredness and selfishness of many individuals across the professions who value their personal success more than the core principles of their professions and service to their communities.

The status of our institutions of higher learning has steadily declined. Indeed, in recent years, a significant proportion of the population of the United States—and, notably, a majority of those who identify as Republican—see our universities as detrimental to the national interest.

The knowledge of experts has also ceased to be valued. On both the left (with its postmodern excesses) and right (with its antagonism to expertise and to institutions of higher learning that it regards as politically torqued), there is skepticism about truth, if not outright embracement of fake news, alternative facts, and alternative realities. Needless to say, disinterested universities and professions cannot survive when respect, veneration—indeed, love—of truth have been abandoned.

I wish I could say that this lamentable state of affairs is just an American phenomenon, but it is not. We see the same signs all over the world. In recent years, we have all witnessed the rise and hegemony of digital technology. As early as 2005, I was skeptical about the "democratic promise" of the new media. From that moment on, we have sought in our research to document and detail the dangers as well as the positive potentials inherent in these new media.

That said, I am not a techno-determinist. The status of professions in the developed world was being undermined well before the advent of the internet, the World Wide Web, and social media. In the United States, institutions of higher learning reached their peak in the 1960s and never fully recovered from the political and sociocultural chaos of the late 1960s and early seventies, while states' investment in their public colleges and universities was challenged by punitive voting propositions even before Steve Jobs and Bill Gates launched their enterprises. As for the disinterested pursuit of knowledge, postmodern scholars, particularly from France, had little use for concepts of truth—or beauty or goodness—long before Fox News was a glint in Roger Ailes's eyes.

The only way to preserve and resurrect the professions, universities, and disinterested scholarship is for those of us who believe in them to work as hard as we can to make them as exemplary, admirable, and attainable as possible. Toward these ends, my colleagues and I at Harvard Project Zero, a research program focused on the arts and humanities, have, for more than twenty-five years, tried to understand what we call "good work" in the professions—work that is at once excellent in quality, personally engaging, and carried out in an ethical manner. In so doing, we have sought to be careful scholars, indicating what we have found, correcting the record when we can, and presenting and publishing our findings in reputable truth-valuing outlets. For the last eight years, we have also studied institutions of higher education. In our writings, we seek to indicate the pressures that these institutions confront and to highlight those schools and practices that exemplify the highest standards and expectations. Finally, over the years, we have developed many curricular interventions designed to engender "good work" and "good citizenship" in K–12 and college students.

In no way do we reject the new technologies. We believe that, like all tools, they can be put to various uses. It's up to us to work to help them support the values we cherish. Of course, by ourselves we can't bring about a reassertion of exemplary quality in the professions, across colleges and universities, and in the propositions and conclusions put forth by scholars. But only if those of us who do believe in these sacred human inventions do our utmost will we have any chance of bringing about a future in which we hope that those who come after us will have the opportunity to live and thrive.

PART 5

TECHNOLOGY AND RACE

The New Jim Code

RUHA BENJAMIN

Ruha Benjamin is associate professor of African American Studies at Princeton University. She is also the founder of the JUST DATA Lab, editor of Captivating Technology, *and author of* People's Science *and* Race after Technology.

When it comes to search engines such as Google, it turns out that online tools, like racist robots, reproduce the biases that persist in the social world. They are, after all, programmed using algorithms that are constantly updated on the basis of human behavior and are learning and replicating the technology of race, expressed in the many different associations that the users make. This issue came to light in 2016, when some users searched the phrase "three Black teenagers" and were presented with criminal mug shots. Then when they changed the phrase to "three white teenagers," users were presented with photos of smiling, go-lucky youths; and a search for "three Asian teenagers" presented images of scantily clad girls and women. Taken together, these images reflect and reinforce popular stereotypes of Black criminality, white innocence, and Asian women's sexualization that underpin much more lethal and systemic forms of punishment, privilege, and fetishism respectively. The original viral video that sparked the controversy raised the question "Is Google being racist?," followed by a number of analysts who sought to explain how these results were produced: "The idea here is that computers, unlike people, can't be racist but we're increasingly learning that they do in fact take after their makers . . . Some experts believe that this problem might stem from the hidden biases in the massive piles of data that algorithms process as they learn to recognize patterns . . . reproducing our worst values." According to the company, Google itself uses "over 200 unique signals or 'clues' that make it possible to guess what you might be looking for." Or, as one observer put it, "the short answer to why Google's algorithm returns racist results is that society is racist."

However, this does not mean that we have to wait for a social utopia to float down from the clouds before expecting companies to take action. They are already able to optimize online content in ways that mitigate bias. Today, if you look up the keywords in *Algorithms of Oppression* author Safiya Noble's iconic example, the phrase "Black girls" yields images of Black Girls Code founder Kimberly Bryant and #MeToo founder Tarana Burke, along with images of organizations like Black Girls Rock! (an awards show) and Black Girls Run! (a wellness movement). The technical capacity was always there, but social awareness and incentives to ensure fair representation online were lacking. As Noble reports, the pornography industry has billions of dollars to throw at companies in order to optimize content, so advertising cannot continue to be the primary driver of online content. Perhaps Donald Knuth's proverbial warning is true: "Premature optimization is the root of all evil."

And so the struggle to democratize information gateways continues. A number of other examples illustrate algorithmic discrimination as an ongoing problem. When a graduate student searched for "unprofessional hairstyles for work," she was shown photos of

Black women; when she changed the search to "professional hairstyles for work," she was presented with photos of white women. Men are shown ads for high-income jobs much more frequently than are women, and tutoring for what is known in the United States as the Scholastic Aptitude Test (SAT) is priced more highly for customers in neighborhoods with a higher density of Asian residents: "From retail to real estate, from employment to criminal justice, the use of data mining, scoring and predictive software . . . is proliferating . . . [And] when software makes decisions based on data, like a person's zip code, it can reflect, or even amplify, the results of historical or institutional discrimination."

In 2015 Google Photos came under fire because its auto-labeling software tagged two Black friends as "gorillas"—a racist depiction that goes back for centuries, formalized through scientific racism and the association of Black people with simians in the Great Chain of Being. It found its modern incarnation in cartoons of former First Lady Michelle Obama and Roseanne Barr's racist tweets against Valerie Jarrett and was resuscitated in algorithms that codify representations used for generations to denigrate people of African descent. This form of machine bias extends beyond racialized labels, to the very exercise of racist judgments.

Further concerns may arise as AI is given agency in our society. The practice of codifying existing social prejudices into a technical system is even harder to detect when the stated purpose of a particular technology is to override human prejudice.

In October 2018, Amazon scrapped an AI recruitment tool when it realized that the algorithm was discriminating against women. The system ranked applicants on a score of 1 to 5; it was built using primarily the resumes of men over a ten-year period and downgraded applications that listed women's colleges or terms such as "women's chess club." But even after programmers edited the algorithm to make it remain "gender neutral" to these obvious words, Amazon worried that "the machines would devise other ways of sorting candidates that proved discriminatory." They rightly understood that neutrality is no safeguard against discriminatory design. In fact, given tech industry demographics, the training data were likely much more imbalanced by race than by gender, so it is probable that the AI's racial bias was even stronger than the reported gender bias.

Some job seekers are already developing ways to subvert the system by trading answers to employers' tests and by creating fake applications as informal audits of their own. In fact one HR employee for a major company recommends "slipping the words 'Oxford' or 'Cambridge' into a CV in invisible white text, to pass the automated screening." In terms of a more collective response, a federation of trade unions called UNI Global Union has developed "a charter of digital rights for workers touching on automated and AI-based decisions, to be included in bargaining agreements."

The danger of New Jim Code impartiality is the neglect of ongoing inequity perpetuated by colorblind designs. In this context, algorithms may not be just a veneer that covers historical fault lines. They also seem to be streamlining discrimination–making it easier to sift, sort, and justify why tomorrow's workforce continues to be racially stratified. Algorithmic neutrality reproduces algorithmically sustained discrimination.

Adapted from Race after Technology: Abolitionist Tools for the New Jim Code *(Cambridge: Polity Press, 2019).*

Closing the Digital Divide

GEOFFREY CANADA

From 1990 to 2014, Geoffrey Canada served as president and chief executive officer of the Harlem Children's Zone, which the New York Times *called "one of the most ambitious social-policy experiments of our time." In 2011, Canada was named to the* Time *100 list of the most influential people in the world, and in 2014, he was named one of* Fortune *magazine's fifty greatest leaders in the world. He is the author of* Fist, Stick, Knife, Gun: A Personal History of Violence.

Back in the late fifties and early sixties, when I was growing up in the South Bronx, my mother was the only person of color I knew who invested in a set of encyclopedias. At that time, encyclopedias supposedly contained all the information in the known world. A set cost an extraordinary amount of money, and we couldn't really afford it, but my mother paid a little bit for it every month. That was the investment she made in her kids' education.

Encyclopedias were compiled alphabetically, and my ambition as a child was to start at A and go all the way through Z. By reading about China and Africa and so many other places and events, I gained a view of the world that propelled me, as well as my brothers, out of the urban ghetto and prepared us for a different academic trajectory.

Today, of course, some kids have digital devices and connectivity that give them access to all the known information in the world, while other kids have nothing but a TV. But the fact is, today, digital connectivity is as essential as having clean drinking water. It's as necessary to education as an up-to-date textbook. In the past, when children of color were given textbooks that were thirty years old, we passed laws to try and stop that kind of inequality. But we face the same challenges today because of lack of access and connectivity. There's a whole bunch of children in this country who have the world's library at their fingertips, but as many as 30 percent of them do not.

That digital divide became very obvious when schools across America closed due to the COVID-19 pandemic and teaching and learning went online. A 2020 study by Common Sense Media showed that at least a third of K–12 students did not have internet or digital devices at home—the essential tools they needed for remote learning. Most of those children live in Black, rural, Latinx, or Native American communities. Without remote learning tools, those kids face significant learning loss. In minority communities, students may be set back as much as a year. So kids who were in the fourth grade before the pandemic and assume they'll be entering fifth grade will really be entering at the fourth grade level when the next school year begins. It's a disaster. It's unconscionable and un-American. We all have to be thinking about these issues in a systematic way.

Digital devices—and adequate connectivity—are now basic necessities. Without them, kids don't have a fighting chance. There have to be very inexpensive but high-quality basic technology options that everybody has the ability to afford. That includes connectivity that allows for streaming—not for watching movies, but to enable students' participation in math, science, English, and history classes that are being taught remotely. If your device constantly freezes while you're trying to learn algebra, it creates an impossible situation for both students and teachers.

At the Harlem Children's Zone—a child-oriented educational and parenting nonprofit I've worked with for thirty years—we've made a huge investment in technology. Before the mayor closed New York City's schools in March 2020, we prepared a plan and surveyed all of our students to find out who had connectivity. We gave hotspots to those who didn't have it, and we distributed more than a thousand devices, including headphones, to make sure that every child in every family had access to distance learning.

Today, closing the achievement gap means closing the digital divide. Every family needs to get connected and understand the educational power of digital technology. Beyond that, parents need online connection these days to do the simplest, most important things–from applying for jobs or unemployment to accessing telehealth care during a pandemic emergency. But too many people can't pay the high cost of phone plans. Are they going to spend the little money they have on phones or food for their families? As a result, they have their services cut off, but they have no free wireless alternatives. Without online access, we're taking a system that's already unequal and making it more so.

Online connection has to be universal, from the inner cities to rural Appalachia and Mississippi. Without educational equity, we will never have an equal society. We have got to prepare our children, give them the tools they need to access a good education–in the classroom and remotely–and make sure they learn.

The Newest Jim Crow

MICHELLE ALEXANDER

Michelle Alexander is a civil rights lawyer, advocate, legal scholar, and author of The New Jim Crow: Mass Incarceration in the Age of Color Blindness.

In 2018, Michigan became the first state in the Midwest to legalize marijuana, Florida restored the vote to over 1.4 million people with felony convictions, and Louisiana passed a constitutional amendment requiring unanimous jury verdicts in felony trials. These are the latest examples of the astonishing progress that has been made in the last several years on a wide range of criminal justice issues. Since 2010, when I published *The New Jim Crow*—which argued that a system of legal discrimination and segregation had been born again in this country because of the war on drugs and mass incarceration—there have been significant changes to drug policy, sentencing, and reentry, including "ban the box" initiatives aimed at eliminating barriers to employment for formerly incarcerated people.

This progress is unquestionably good news, but there are warning signs blinking brightly. Many of the current reform efforts contain the seeds of the next generation of racial and social control, a system of *e-carceration* that may prove more dangerous and more difficult to challenge than the one we hope to leave behind.

Bail reform is a case in point. Thanks in part to new laws and policies—as well as actions like the mass bailout of people locked in New York City jails that's underway—the unconscionable practice of cash bail is finally coming to an end. In August 2018, California became the first state to decide to get rid of its cash bail system; last year, New Jersey virtually eliminated the use of money bonds.

Increasingly, computer algorithms are helping to determine who should be caged and who should be set "free." Freedom—even when it's granted, it turns out—isn't really free. Under new policies in New Jersey, New York, and beyond, "risk assessment" algorithms recommend to judges whether a person who's been arrested should be released. These advanced mathematical models—or "weapons of math destruction" as data scientist Cathy O'Neil calls them—appear colorblind on the surface, but they are based on factors that are not only highly correlated with race and class, but are also significantly influenced by pervasive bias in the criminal justice system.

As O'Neil explains, "It's tempting to believe that computers will be neutral and objective, but algorithms are nothing more than opinions embedded in mathematics." Challenging these biased algorithms may be more difficult than challenging discrimination by the police, prosecutors, and judges. Many algorithms are fiercely guarded corporate secrets. Those that are transparent—you can actually read the code—lack a public audit, so it's impossible to know how much more often they fail for people of color.

Even if you're lucky enough to be set "free" from a brick-and-mortar jail thanks to a computer algorithm, an expensive monitoring device likely will be shackled to your ankle—a GPS tracking device provided by a private company that may charge you around $300 per month, an involuntary leasing fee. Your permitted zones of movement may make it difficult or impossible to get or keep a job, attend school, care for your kids, or visit family members. You're effectively sentenced to an open-air digital prison, one that may not extend beyond your house, your block, or your neighborhood. One false step (or one malfunction of the GPS tracking device) will bring cops to your front door, your workplace, or wherever they find you, and snatch you right back to jail.

Who benefits from this? Private corporations. According to a report released last month by the Center for Media Justice, four large corporations–including the GEO Group, one of the largest private prison companies–have most of the private contracts to provide electronic monitoring for people on parole in some thirty states, giving them a combined annual revenue of more than $200 million just for e-monitoring. Companies that earned millions on contracts to run or serve prisons have, in an era of prison restructuring, begun to shift their business model to add electronic surveillance and monitoring of the same population. Even if old-fashioned prisons fade away, the profit margins of these companies will widen so long as growing numbers of people find themselves subject to perpetual criminalization, surveillance, monitoring, and control.

Who loses? Nearly everyone. A 2018 analysis by a Brookings Institution fellow found that "efforts to reduce recidivism through intensive supervision are not working." Reducing the requirements and burdens of community supervision, so that people can more easily hold jobs, care for children, and escape the stigma of criminality "would be a good first step toward breaking the vicious incarceration cycle," the report said.

Many reformers rightly point out that an ankle bracelet is preferable to a prison cell. Yet I find it difficult to call this progress. As I see it, digital prisons are to mass incarceration what Jim Crow was to slavery.

If you asked slaves if they would rather live with their families and raise their own children, albeit subject to "whites only signs," legal discrimination, and Jim Crow segregation, they'd almost certainly say: I'll take Jim Crow. By the same token, if you ask people in prison whether they'd rather live with their families and raise their children, albeit with nearly constant digital surveillance and monitoring, they'd almost certainly say: I'll take the electronic monitor. I would too. But hopefully we can now see that Jim Crow was a less restrictive form of racial and social control, not a real alternative to racial caste systems. Similarly, if the goal is to end mass incarceration and mass criminalization, digital prisons are not an answer. They're just another way of posing the question.

Some insist that e-carceration is "a step in the right direction." But where are we going with this? A growing number of scholars and activists predict that "e-gentrification" is where we're headed as entire communities become trapped in digital prisons that keep them locked out of neighborhoods where jobs and opportunity can be found.

If that scenario sounds far-fetched, keep in mind that mass incarceration itself was unimaginable just forty years ago and that it was born partly out of well-intentioned reforms—chief among them mandatory sentencing laws that liberal proponents predicted would reduce racial disparities in sentencing. While those laws may have looked good on paper, they were passed within a political climate that was overwhelmingly hostile and punitive toward poor people and people of color, resulting in a prison-building boom, an increase in racial and class disparities in sentencing, and a quintupling of the incarcerated population.

Fortunately, a growing number of advocates are organizing to ensure that important reforms, such as ending cash bail, are not replaced with systems that view poor people and people of color as little more than commodities to be bought, sold, evaluated, and managed for profit. In July 2018, more than one hundred civil rights, faith, labor, legal, and data science groups released a shared statement of concerns regarding the use of pretrial risk assessment instruments; numerous bail reform groups, such as Chicago Community Bond Fund, actively oppose the expansion of e-carceration.

If our goal is not a better system of mass criminalization, but instead the creation of safe, caring, thriving communities, then we ought to be heavily investing in quality schools, job creation, drug treatment, and mental health care in the least advantaged communities rather than pouring billions into their high-tech management and control. Fifty years ago, the Rev. Dr. Martin Luther King Jr. warned that "when machines and computers, profit motives and property rights are considered more important than people, the giant triplets of racism, extreme materialism and militarism are incapable of being conquered." We failed to heed his warning back then. Will we make a different choice today?

From the New York Times.

Technology, Inclusiveness, Structural Racism, and Silicon Valley

THEODORE M. SHAW

Theodore M. Shaw is the Julius L. Chambers Distinguished Professor of Law and director of the Center for Civil Rights at the University of North Carolina School of Law.

One summer night in 1989, under a star-filled sky, on the grounds of the Great Windsor Park in England—far from the work I did as a civil rights lawyer in the United States—I had a revelation. While participating in a seminar on global interdependence, it occurred to me that the role of Black Americans in the United States—the sole purpose for which they had been involuntarily brought to America—had disappeared. African Americans, brought to the Western Hemisphere four centuries earlier as slave labor, no longer served that purpose or its successor purpose, as the servant class. After the civil rights movement won legal rights to equal educational, employment, and other opportunities, Black Americans did not automatically become equal participants in the US economy. Collectively, because of disadvantages accumulated over hundreds of years, Black Americans were economically superfluous.

In the aftermath of the civil rights movement, the United States transitioned from a postindustrial economy to an information-based economy. Many of the jobs for which Black workers fought were becoming obsolete or moving overseas. Educational qualifications for the new economy grew beyond reading literacy to include computer literacy. In the new economy, the opportunity and wealth gaps between white and Black Americans grew even wider. A new wave of entrepreneurs accumulated enormous wealth in Silicon Valley and other tech centers. Most Black people were left behind or excluded from the new economy. Big tech companies in Silicon Valley employ abysmally small percentages of Black and brown employees, hovering in the low single digits. While they pay lip service to diversity, they are doing little to achieve it.

The tech industry and Silicon Valley did not exist during the age of legally enforced segregation. They may not feel any obligation to address structural racism and inequality in America. Yet it is impossible to think about how technology is reshaping democracy and our lives without thinking about the impact of media and technology on race and society.

In the 1960s, media and technology played an important role in advancing the cause of civil rights. The strategy of nonviolent resistance depended, in large part, on exposing the moral bankruptcy of segregation and racial violence. In the Birmingham campaign, images of firehoses turned on peaceful protesters and police dogs ripping at their clothes and flesh were broadcast on the evening news and around the world. These images helped persuade President Johnson and Congress to enact civil rights legislation. The March on Washington–and the image of a young and powerfully eloquent Martin Luther King Jr. challenging America to live up to its ideals–is emblazoned in the minds of generations of Americans, most of whom were not alive to see it in real time. Technology, however, made these moments eternally part of our national psyche.

In our own time, ordinary individuals with cell phones are documenting police brutality and the deaths of unarmed Black people in ways that have illuminated patterns and practices that for so long were denied and ignored, as if they did not matter. In 2020, the deaths of George Floyd, Ahmaud Arbery, and Rayshard Brooks, captured by technology, gave rise to an organic movement calling for a reexamination of structural racism and policing. The images of the way in which a police officer–sworn to protect and defend life and the Constitution–casually and callously killed a nonthreatening, compliant Black man, pleading for his life and for his mother, cannot be unseen.

Ironically, technology–designed and produced by companies that are failing miserably to be inclusive in their hiring practices–is finally making it impossible for many white people to deny the way in which racism and white supremacy are structurally baked into American institutions and individuals. Silicon Valley needs to do better. Technology has created a new economy that is leaving Black and brown workers further behind. It has also exposed American racism and shown the world that, at long last, America has to become better.

Technically Racist

SHAUN R. HARPER

A provost professor in the Marshall School of Business at the University of Southern California, Shaun R. Harper is the Clifford and Betty Allen Chair in Urban Leadership and executive director of the USC Race and Equity Center. He has published twelve books and consulted with more than three hundred companies and institutions on equity, diversity, and inclusion. Dr. Harper also is an editor-at-large of Time *magazine.*

Most people do not think of themselves as racist. When told they did something that may have been experienced as racism, they typically insist that they have been misunderstood. Resistance and defensiveness are common responses. Although I have never worked in the tech industry, I am guessing that most of its professionals do not think of their field as racist. Some would likely say that tech is not where it should be in terms of the racial composition of the workforce. But racist? No. "Sure, we have a severe shortage of Black and Latino employees, but there's no actual racism here." That is what I suspect most in Silicon Valley and elsewhere would say about their industry in general, and their companies in particular. A response like this signals a limited understanding of what is *technically* racist.

Through their participation in protests during the month of June 2020, activists and their supporters around the world raised public consciousness about structural and systemic racism. While the police killings of George Floyd and Breonna Taylor catalyzed these uprisings, the movement quickly revealed other manifestations of racism, anti-Blackness, and white supremacy. During this time, several courageous Black employees across industries, including tech, wrote open letters to their colleagues about their firsthand encounters with racism in their workplaces. Others did so in company-wide town hall forums on race that had been quickly thrown together without strategy. This surprised white executives–I know so, because too many of them called me for advice. Although I was meeting these leaders for the first time, it was obvious to me (regardless of whether we were videoconferencing or talking on the phone) that they were distressed and afraid. None told me that they thought of themselves as racist; it was disorienting that they could be personally implicated in the horrifying stories their Black colleagues were sharing. It was also shocking to most executives that so many Black employees and other colleagues of color were suddenly coming forward with detailed examples of racism they had long experienced at work. Why were these leaders so surprised? Because they had not previously thought of what was being expressed in open letters and company forums as technically racist.

One aspect of the tech industry's longstanding diversity problem is well documented and widely known: the shortage of Black and Latino employees. While this is partially explained by pipeline shortages, it is also attributable to laziness, excuse making, and racial bias in recruitment and hiring. It is important to note that racism in tech is about much more than the number of employees of color. In their open letters, company forums, social media posts, press interviews, and elsewhere, Black employees in tech and other industries described a range of realities that, technically, are racist.

For example, there is racial stratification in their workplaces. Black professionals tend to be in the least compensated, lowest-paid roles in companies. They are also almost always severely underrepresented in leadership positions. Their managers invest more heavily and more routinely in the career ascension of their white counterparts. Furthermore, Black employees who have participated in my workplace climate studies over the past decade have repeatedly told me that they are passed over time and time again for promotions, while their less talented, less accomplished white coworkers are groomed for and accelerated to higher-paying, more senior positions. Technically, this is racist.

Black professionals in my research also often characterize their workplace climates as racist. Being mistaken for another Black colleague to whom they bear no resemblance, technically, is racist. So, too, is having to constantly fight for more Black workers to be hired and promoted; having to beg for more resources for the Black employee resource group; and having to justify programming during Black History Month. Asking a Black colleague to speak on behalf of all Black people, all people of color, or all poor people, technically, is racist. Touching a Black person's hair without permission, technically, is racist. Having one's bright idea unacknowledged while, moments later, a white colleague is praised for having the same idea, technically, is racist. Assuming every Black man likes hip hop music, and speaking slang only to him, is technically racist. And asking a Black coworker for special permission to use the n-word is definitely racist. These are only some experiences that many Black employees have in tech and other companies.

Executives and others in the tech industry tend to think about racism in narrow terms—as denying interviews to applicants with ethnic-sounding names, refusing to hire a person simply because they are Latino, calling people the n-word and other racial epithets, or burning a cross on a Black family's lawn. Those indeed are racist acts. But technically, racism is also the other aforementioned experiences that employees of color describe to me in my research studies. Moreover, it is clear to me that leaders were surprised by stories shared in June 2020 because they had not previously taken time to talk with Black employees. They had not relied on data about the workplace racial climate to ascertain what their colleagues of color were feeling and experiencing. They had invested too little effort into identifying, recruiting, hiring, promoting, and retaining Black employees. That is why they were so surprised and unprepared to respond to the uprisings that occurred within their companies after protests against anti-Black racism erupted globally.

Being on the right side of history requires not only denouncing racism and declaring that Black lives matter. It demands confronting and dismantling racism in all its forms, including workplace racism. It requires formally assessing the racial climates of tech companies, instead of presuming that there is no racism in those companies and that all employees, across all racial groups, feel included, respected, and treated fairly. Two or three broad, generic questions on their annual employee experience questionnaires will not offer sufficient insights. Surveys and interviews focused deeply and specifically on the racial climate are needed. Tech leaders must then take meaningful strategic action in response to problems that their workplace climate assessments reveal.

It is not enough to hire a chief diversity and inclusion officer, who has no serious budget and too few staff members; to allow employees to start resource groups with no money or power to enact systemic change in the company; or to do a one-time workshop, focused broadly on inclusion, that only one-tenth of employees attends. Those actions will do nothing to place and keep tech companies on the right side of history. And most certainly, the annual ethnic food potluck event will not fix a company's deeply entrenched, multifaceted racial problems.

Instead, strategy and resources are required, as well as race-centric professional learning experiences for employees at all levels, including executives and managers, and holding oneself and everyone inside the company accountable for racial equity. Tech leaders who wish to be on the right side of history must go beyond saying racism is bad. They must be courageous and intentional about demonstrating their commitment to understanding, routinely assessing, honestly naming, and strategically dismantling it throughout their companies. That, technically, is what antiracism is.

PART 6

DOING GOOD, NOT EVIL

Inside Cult 2.0

RENÉE DiRESTA

Renée DiResta is the technical research manager at Stanford Internet Observatory. She studies influence operations and computational propaganda in the context of pseudoscience conspiracies, terrorist activity, and state-sponsored information warfare, and she has advised Congress, the State Department, and other academic, civil society, and business organizations on the topic.

We used to worry about filter bubbles—search or social results that inadvertently trapped users in a certain sphere of *information.* But in today's era of hyperpartisan politics, fragmented media, and low trust in institutions—in which social network groups have become a primary social structure—we've entered the realm of *bespoke realities.* People have become enmeshed in cult-like online communities that operate with their own media, facts, authority figures, and norms. Some behave quite a lot like a cult—Cult 2.0.

Some common pathways are reported by people who fall into and then leave these highly insular communities: They usually report that their initial exposure started with a question, and that in response, a search engine took them to content that they found compelling. They engaged with that content and then found more. They joined a few groups, and soon a social network's recommendation engine sent them to others. They alienated old IRL friends but made new ones in various conspiratorial groups, where they jointly investigated obscured truths and hidden connections, built a trusted community, and recruited other people.

"When you met an ignorant nonbeliever, you sent them YouTube videos of excessively protracted contrails and told them things like: 'Look at the sky! It's obvious!'" Stephanie Wittis, a self-described former chemtrails and Illuminati conspiracy believer, told *Vice*. "You don't even go into detail about the matter or the technical inconsistencies, you just give them any explanation that sounds reasonable, cohesive, and informed—in a word, scientific. And then you give them the time to think about it."

We have reached the point when even the Federal Bureau of Investigation (FBI) has come to believe that online conspiracy theories present a risk of escalating to domestic terrorism. There are, unfortunately, plenty of examples.

In 2018, investigations of Cesar Sayoc, dubbed the "MAGA-bomber," uncovered his participation in conspiratorial Facebook groups. The Tree of Life synagogue shooter was active in communities on several social networks that believed that Jews were engineering mass migration to the United States. The Comet Ping Pong "Pizzagate" shooter belonged to online Pizzagate communities. A QAnon deep-state conspiracy adherent held a one-man standoff at the Hoover Dam. Another QAnon supporter was arrested occupying a Cemex cement factory, claiming that he had knowledge that Cemex was secretly assisting in child trafficking.

"The easiest way to radicalize someone is to permanently warp their view of reality," says Mike Caulfield, head of the American Association of State Colleges and Universities Digital Polarization Initiative. "It's not just confirmation bias. . . . We see people moving step by step into alternate realities. They start off questioning and then they're led down the path." The process is similar to cult-recruitment tactics of the pre-internet era, in which recruits are targeted and then increasingly isolated from the noncult world. In the era of online recommendation algorithms, however, the path has a significant number of possible forks: An antivaccine conspiracy believer is likely to receive suggestions for nonvaccine conspiracies that still involve elements of distrust of government (say, Pizzagate and QAnon) or elements of distrust in accepted science (flat earth, chemtrails groups). By keying off a set of signals derived from what's known as "collaborative filtering," recommendation engines may inadvertently act as a kind of conspiracy correlation matrix that keys off of a complex array of signals—including the interests of other people who share similarities with the targeted user.

While the specific narrative may vary, the path ultimately leads to closed online communities. Members are unlikely to have real-world connections but are bound by shared beliefs. Some of these groups, such as the QAnon Facebook communities, number in the tens of thousands, and there are dozens of them. "What a movement such as QAnon has going for it, and why it will catch on like wildfire, is that it makes people feel connected to something important that other people don't yet know about," says cult expert Rachel Bernstein, who specializes in recovery therapy. "All cults will provide this feeling of being special."

The idea that "more speech" will counter these ideas fundamentally misunderstands the dynamic of these online spaces: Everyone else in the group is also a true believer. The articles they share and the memes they create reinforce their worldview. There is no counterspeech. Inside Cult 2.0, dissent is likely to be met with hostility and expulsion from the group, possibly accompanied by doxing and sustained harassment. No one in the online community is going to report indications of radicalization to the platforms' Trust and Safety portals.

Digital researchers and product designers have a lot to learn from looking at cult deprogramming and counter-radicalization work done by psychologists. "When people get involved in a movement, collectively, what they're saying is they want to be connected to each other," notes Bernstein. "They want to have exclusive access to secret information other people don't have, information they believe the powers that be are keeping from the masses, because it makes them feel protected and empowered. They're a step ahead of those in society who remain willfully blind. This creates a feeling similar to a drug—it's its own high."

This conviction largely inures members to correction, which is a problem for the fact-checking initiatives that platforms have attempted. After Facebook tried to fact-check misinformation, researchers found—counterintuitively—that people doubled down and shared the article more when it was disputed. "*They don't want you to know*," readers claimed, alleging that Facebook was trying to censor controversial knowledge. YouTube tried fact-checking as well. In late 2018, it began adding Wikipedia links to videos featuring conspiracy theories. That effort, too, appears to have had little impact. For YouTube, in particular, the challenge is compounded by the fact that its recommendation system incentivizes conspiratorial content. As Alexis Madrigal writes in the *Atlantic*, "It's not only that conspiracy content made YouTube viewers more prone to believe conspiracies. It's that the economics and illusions of content production on YouTube itself made conspiracy content more likely to be created and viewed. And these forces have reinforced each other for years, hardening them against the forms of control that YouTube can exert."

What does work to counter indoctrination in real life? One-on-one interventions and deradicalization messages from people within these trusted networks. Since those solutions are very difficult to scale online, experts are looking to prevent people from becoming radicalized in the first place—either by inoculating communities or attempting to intercede as early as possible in their radicalization process. These approaches, however, require companies to fundamentally shift their recommendation process and redirect certain users *away* from what they wish to see. They require the platforms to challenge their own search engines—and perhaps to make judgment calls about the potential harm that comes from certain types of content.

Previously, the platforms have only drawn lines and undertaken thorny evaluations after extensive public pressure and government pleas, such as in cases of explicit terrorist radicalization. To counter violent terrorist threats like ISIS, YouTube ran a program called the Redirect Method, which used ad placement to suggest counter-channels to those searching for ISIS propaganda. Reddit banned several QAnon subreddits on the grounds that they were inciting violence and harassment. Instagram and Pinterest opted not to return search results for queries it deemed detrimental. Beginning in mid-2019, the platforms also began to take action around conspiracy groups that posed a risk of physical harm to adherents or to public health. Cancer quackery, bleach treatments for autism, and antivaccine conspiracy–focused groups were pulled from the recommendation engines—though, in most cases, the content and communities still exist on the platforms and can be found via a bit of searching.

Companies are reluctant to apply these interventions and nudges when conspiratorial indoctrination is less obvious because of valid concerns about the impact on free expression. This is compounded by the fact that we have a minimal amount of data about online radicalization. Only the social platforms themselves have full visibility into what is presented to certain groups of people and in what order. But online radicalization is now a factor in many destructive and egregious crimes, and the need to understand it is gaining urgency. Most of what is understood about the process comes from experiments that attempt to simulate what YouTube is recommending; from collections of anecdotes from Facebook-, Amazon-, and Twitter-focused investigative journalism; from personal testimonies from those who have gone down the rabbit hole; and from survey studies with small sample sizes. Some of the findings conflict. It's hard to isolate the impact of algorithms in a complex system that also has to account for human preferences, offline influences, and ways that people seek out information.

Ultimately, however, this is an area of growing concern that we need to understand. As the FBI wrote in its May 2019 memo, "conspiracy theory–driven domestic extremists" are a growing threat. The agency "assesses these conspiracy theories very likely will emerge, spread, and evolve in the modern information marketplace, occasionally driving both groups and individual extremists to carry out criminal or violent acts."

The promise of social networks was that they would bring people together, wherever they were. Amoral algorithms that inadvertently turn online groups into cults have shown the potential for unintended consequences in the era of bespoke realities. The systems that mediate our ways of connecting require more vigilance.

Transforming the Attention Economy

TRISTAN HARRIS

A former design ethicist at Google, Tristan Harris is cofounder and president of the Center for Humane Technology and cohost of the podcast Your Undivided Attention.

A decade ago, Edward O. Wilson—a Harvard professor and renowned father of sociobiology—observed that "the real problem of humanity is the following: We have Paleolithic emotions, medieval institutions, and godlike technology." Since Wilson made that statement, technology's godlike powers have grown exponentially, while the ancient impulses, biases, and emotions of our brains have remained the same.

While it is fashionable to think that the number-one harm of technology is that it has failed to protect our privacy and our data, critics often overlook the potentially more consequential harms that emerge from a different problem: *technology outpacing the limitations of our brains.* For example: Digital addiction, political polarization, election manipulation, teen depression, the outrage-ification of politics, the breakdown of truth, and the rise of microcelebrity vanity culture all emerge not from technology disrespecting our privacy but from technology being deployed in a multibillion-dollar battle to hijack our brains and capture our attention. That's because the more time we spend on their platforms watching, clicking, liking, and sharing posts and videos, the more ads we see and the more billions of dollars are poured into the balance sheets of tech companies.

The very harms we want to avoid are directly linked to hijacking our brains for profit. Addiction is more profitable than casual or occasional use. Lonely screen time is more profitable than real-life time laughing with friends over dinner tables. Transforming a generation of teenagers into attention seekers with low self-esteem, addicted to how many followers or likes they have, is more profitable than a generation of teenagers who are self-confident about their personhoods. News feeds that force-deliver us personalized affirmations are more profitable than news feeds that challenge our beliefs, shredding reality into 2.7 billion *Truman Shows* and making it impossible to find common ground.

YouTube, Snapchat, Instagram, and Facebook's feeds and recommendations have figured out that hijacking our attention is as simple as out-predicting what will trigger our Paleolithic emotions and biases. The reality is, harms like addiction, social isolation, teen depression, dangerous polarization, and constant distraction are not separate issues. They are an interconnected system of harms directly emerging from the social media platform's toxic business model.

This results in a backward step in society's evolution to what we call "human downgrading"—an ever-more-aggressive race to capture and manipulate human attention and erode our capacity for meaningful choice. We are being mined like uranium ore, all in the pursuit of enormous profits.

What Happens When the Supercomputer Aims at You

Let's take an example. Think of a moment when you opened a YouTube video sent to you by a friend. Right before you opened it, you thought to yourself, "I'm going to watch this one video, then go back to my other work." After then waking up from two hours of a hypnotic-like trance of binge-watching, you think to yourself, "What just happened?" One answer is that you should have had more self-control. Sounds reasonable. But that explanation masks the real magic trick: When you hit play, you had Google's most powerful supercomputer pointed at your brain. And it did not have your best interests at the forefront of its algorithmic priorities.

Their supercomputer's goal is to figure out how to maximize how long you stay on the site, using their psychographically targeted recommendations. To generate the video recommendations most likely to keep you watching, imagine YouTube waking up an avatar voodoo doll–like version of yourself inside a server. This avatar predictive model is based on all your clicks, likes, everything you've watched, places you regularly travel to, and how your behavior compares to other avatars. Over time, these features learn from your data and, in a strange hall of mirrors, make the avatar look and act more and more like you; it becomes your cyber twin. Its simulated behaviors and responses increasingly mirror how you would respond in real life. YouTube doesn't have to run tests on you because it can just run test scenarios on your avatar to determine which videos will keep you there the longest, so that they can keep showing you ads.

As a result of this behind-the-scenes manipulation, YouTube's average watch time is more than sixty minutes a day on mobile—specifically because of what the recommendation engines are putting in front of you. Out of the more than a billion hours of YouTube videos watched daily, 70 percent come from the recommendation system.

Imagine a spectrum ranging from calm videos (Carl Sagan, Walter Cronkite, or David Attenborough) to "crazy town"—conspiracies, aliens, extremism, bigotry, and hatred. No matter where you start, in which direction do you think YouTube will steer you if it wants you to watch more? It will always tilt the floor toward "crazy town." Think of a vast distributed version of the adage "If it bleeds, it leads" inadvertently discovered by soulless, amoral algorithms jacked into more than two billion human brains.

Until 2018, if you were a teen girl and started watching a dieting video, YouTube's algorithms would recommend anorexia videos because those were the things that were better at keeping teen girls' attention. If you watched an educational video on the climate crisis, 50 percent of the YouTube-recommended videos espoused positions opposing scientifically based research. If you watched a NASA moon landing, YouTube recommended flat earth conspiracy videos hundreds of millions of times. YouTube recommended Alex Jones's *InfoWars* conspiracy videos more than fifteen billion times—more than the *combined* monthly traffic to the websites of the *New York Times*, Fox News, and the BBC.

YouTube's engineers have now addressed the flat earth and teen-girl-anorexia recommendations. They have hired tens of thousands of content moderators, whose task is like trying to catch water with a net. Consider that YouTube has created two billion *Truman Shows* in countries with languages that their engineers do not even speak. There are more than twenty-two languages in India, where an election took place in 2018. Do you think YouTube's few thousand moderators caught all the toxic, false-news recommendations happening in all those languages, across millions of videos? Or Facebook, either, for that matter? How successful was Facebook in catching the hate-filled posts and news stories in Myanmar, which has a hundred languages, that were targeted at the Rohingya Muslim minority?

At this point, this much is clear: Tech companies have created a digital Frankenstein they can't control. Automated machines that are unpaid and never get tired or need to sleep are cheaper than paying human moderators or curators. Since Facebook's customer is the advertiser, to whom you the viewer are being delivered like sushi on a platter, anyone who pays the fee can essentially walk into that global psychological control room and say, "I want to microtarget this particular group and fire ideas directly at their minds."

Almost three billion people use Facebook and YouTube collectively each day, which is a psychological footprint bigger than Christianity. That's 25 percent of the world's population, accounting for more than 90 percent of the world's GDP, whose thoughts are shaped by a handful of mostly twenty- to forty-year-old engineers and designers in California. These companies have a titanic amount of influence—more than any government—over people's daily thoughts, beliefs, and information gathering. We expect our government to be accountable, we expect oil or chemical companies to be accountable for their spills—so why are we not holding technology companies accountable to protect our social and cultural well-being?

Just as the complexity and scale of our global problems are increasing—including the threats of climate change, democratic decline, and societal corrosion—our capacity to concentrate, engage in critical thinking, respectfully disagree, solve complex societal challenges, or even have confidence in shared information is declining. This "downgrading" of humanity affects everyone because all of us are trapped in ancient brains surrounded by this attention-extracting environment of technological mining. Slowly, silently, technology is transforming us into a civilization that is maladapted for its own survival.

We Can Change Technology's Path

Here's the good news: We are the only species self-aware enough to identify this existential mismatch between our brains and the technology that is pointed at every aspect of our culture and society.

We can stop it. We have the power, with human choice and shared awareness, to reverse these trends. The question is whether we are genuinely willing to rise to this challenge and to regulate the engagement-driven, social media business model, which threatens to drive our impulses over the cliff. Will we have the wisdom to create technology that is more humane?

We must develop a regenerative technology business model that treats customers as human beings instead of resources to be mined. Technology needs to promote our most natural human strengths and capacities, uplifting our values rather than preying on our minds' vulnerabilities.

I'm optimistic that we can do this. When people started speaking up about these problems—including Roger McNamee; Jaron Lanier; Justin Rosenstein, who invented the Facebook "Like" button; Sandy Parakilas at Facebook; Renée DiResta at Stanford University; Guillaume Chaslot, the ex-YouTube recommendations engineer; and Marc Benioff at Salesforce—things started to change.

In 2014 I gave a TED Talk about why technology needed to transition from an attention economy, in which services compete to extract and maximize the most "time spent," to one in which tech companies compete for "time well spent"—a race to help us live by our values and spend our time *well.* After years of advocacy and public pressure, change finally happened. In 2018, Mark Zuckerberg announced Facebook's big new goal—making sure that time on the platform is "time well spent." The same year, Apple launched "Screen Time" features to help a billion people see and limit their screen time on their phone. So did Google and YouTube. Now, Apple and Google are competing to provide better "digital well-being" experiences for users, and more than a billion devices are running time-well-spent features. These are baby steps, to be sure, but there's promise that, with enough pressure, we can change the currency of competition from attention to a deeper notion of "net positive" benefits to humanity.

The Business Model behind Human Downgrading

The first step in reversing human downgrading is identifying the cause. Advertising was an efficient way to create economic prosperity in the first phase of the internet-based economy. But like coal, it also polluted our environments—personal, cultural, and political—enabling anyone to pay to get access to your mind and to microtarget messages that perfectly persuade and polarize communities and entire populations. "Free" turns out to be the most expensive business model ever created.

The *business model* of these platform companies is the problem, not their amazing technologies. We have to get off this destructive "attention economy" business model and invent a positive alternative.

Advertising itself is not the problem—it's the companies' unlimited appetite to acquire our attention and their incentive to create slight shifts in our identity, behavior, and values with increasing precision. More of our attention means more money for Facebook, YouTube, or Twitter. That's a perverse relationship—like utility companies that profit by urging us to use as much energy as possible, until there's no energy left, or to keep the water running until we drain the reservoir. We know how to solve this kind of problem with regulation. In many states, we've decoupled the amount of money that energy companies earn from how much energy their customers use. It is imperative that we do something similar for the extractive attention economy that is shaping the minds of billions.

To move forward, we need a new set of incentives and disincentives among tech companies that accelerate competition to the top of the brain stem, not the bottom. We need an effort like the mighty one to advance the transition to renewable sources of energy, while also reversing past impacts of climate change. What might that look like?

A New Business Model: The "Attention Utilities"

These businesses are creating the new public infrastructure of the digital age. Search engines, global online portals for news/information/social networking, web-based movies, music and live streaming, GPS-based navigation apps, online commercial platforms and marketplaces, online labor platforms for finding employment, and algorithmic-driven transportation systems have all opened new horizons and caused controversies and challenges. The United States has a long history of enacting "guardrail" regulations when private businesses try to provide crucial infrastructure. When it comes to the "attention economy," we have allowed these businesses to engage in commerce with virtually no rules, regulations, permits, or licenses.

When these companies were fledgling innovators, this laissez-faire policy might have been warranted. Now that these companies are obsessively focused on getting us to use more, more, more, they have scaled into a civilizational threat. We need to create a new business classification—"attention utilities." They have created a public infrastructure that is vital to twenty-first-century progress and need a unique set of rules and regulations guiding their business model and scrutinizing their recommendations, amplification, targeted ads, and content personalization.

To those who see these proposals as well-nigh impossible, consider this: In the 1950s, if you said, "We've got to get off coal," it would have seemed impossible. We didn't have any alternative that would have produced nearly enough energy to support society. Similarly, alternatives to advertising, like subscriptions and micropayments, don't add up yet to a viable business model. But as with renewable-energy technologies, we *can* get to that point if we take the right steps now.

Human downgrading is to culture as atmospheric warming is to the environment. It can be catastrophic if we don't address it. While climate change requires that thousands of businesses in dozens of industries, regulated by hundreds of governments in different countries, all change what they're doing, reversing technology's harms requires that only about a thousand people at a handful of companies—located and regulated by one government in one country—change what they're doing. Tech workers can start raising their voices about the very real and urgent threat associated with human downgrading. Journalists—instead of getting distracted by headline scandals—can shine a light on the systemic problems of tech's parasitic business model and potential solutions.

Voters can demand policies that protect their children and democratic elections by demanding regulation like the Honest Ads Act, which would improve disclosure requirements for online political advertisements. Policymakers can take action, too, by establishing a Department of Attention Economy that would enable them to keep pace with accelerating technology and shift incentives away from the extractive mining of their constituents' attention. Shareholders in technology companies can demand commitments to define a road map away from human-downgrading business models. Venture capitalists can fund the transition to new "regenerative" and humane platforms that aren't built on the unforeseen harms resulting from the commodification of human consciousness and attention. And next-generation entrepreneurs can build new "humane" versions of technology that compete for our trust, not our manipulated attention.

To answer E. O. Wilson's insightful dilemma: Humane technology would *embrace* our Paleolithic emotions, help us *upgrade* our medieval institutions, and help us *gain* the wisdom to carefully wield and slow down godlike technology. We have to do this. Otherwise, like Icarus—who flew too close to the sun on his wings made of wax—we may fall from the sky and drown in a raging, rising sea of our own making.

How Technology Can Humanize Education

SAL KHAN

Sal Khan is the founder and CEO of the nonprofit educational organization Khan Academy, author of the book The One World Schoolhouse: Education Reimagined, *and a teacher at Khan Lab School, a nonprofit mastery-based laboratory school in Mountain View, California.*

While I may be best known as a teacher of math and science, there's a special place in my heart for history. In history, we discuss how civilizations wax and wane because of economic, cultural, and political forces. Students learn about technological revolutions that transform humanity—for example, the arrival of agriculture, the printing press, and the Industrial Revolution.

I believe we are living through a similar transformation right now. We are in the midst of a historical inflection point driven by automation and artificial intelligence. Technology is altering the way we interact with our world and move through our days. In the next twenty years, it's likely that technology will take over everything from driving to grocery checkouts to reading x-ray scans. The coming advancements are on the scale of the Industrial Revolution or even greater.

While technology is likely to make life more convenient for many people, it also raises profound questions. It's likely to disrupt labor markets and exacerbate wealth inequality. What happens to the nearly four million American truck drivers who may lose their jobs? Or the 3.6 million American cashiers? But just as technology is forcing us to ask hard questions about potential job losses, it's also part of the solution. Technology is enabling completely new learning models, and I believe we can use them to help educate our way out of the problem.

If you're skeptical, let's take a look at history. To appreciate the possibility of what we're talking about, consider what it was like to live four hundred years ago in Western Europe, when only about 15 percent of the population knew how to read. I suspect that if you asked someone back then who did know how to read—say, a member of the clergy—"What percentage of the population do you think is capable of reading?" the answer would be, "With a great education system, maybe 20 or 30 percent." Fast-forward to today, and pretty close to 100 percent of the population is capable of reading. That's a huge advance. Similarly, if I were to ask today, "What percentage of the population do you think is capable of mastering calculus, or understanding chemistry, or being able to contribute to cancer research?" many people would likely answer, "With a great education system, maybe 10 or 15 percent."

What if the number of people who could be creative knowledge workers is actually far, far higher? Maybe the reason people have historically struggled with advanced skills has less to do with their innate ability and more to do with knowledge gaps that get wider over multiple years in a fixed-paced school system. For example, if you got 70 percent on a test on basic exponents in seventh grade and 80 percent on a test on negative numbers in eighth grade, it would be very hard for you to keep up with a ninth-grade algebra curriculum that requires mastery in both those topics—no matter how "bright" you may be or how good your teacher is. With traditional educational tools, it's nearly impossible for teachers to address the learning gaps of all thirty students in a class. But new online tools, enabled by technology, can finally personalize learning for individual students at scale. If we do that, I believe the proportion of people capable of participating in the knowledge economy is closer to 80, 90, or even 100 percent.

These new educational software tools, which I and many others are devoting our lives to, are universally accessible and can flex to meet the individual needs of every student. We can help teachers pinpoint their unique needs, identify their learning gaps, and help fill them. Students can progress through personalized lessons, mastering important concepts at the pace that's right for them. The pedagogical term "mastery learning" is one of the most exciting ways to think about education. It's a simple idea—students should master a given concept before being expected to understand a more advanced one. An analysis of nearly three hundred studies on mastery learning found that it has a positive effect on achievement "at all levels and for all subjects." Although the concept of mastery learning has been around for decades, technology is now making it possible at scale.

The data is compelling. In math, for example, as little as one class period per week of mastery-based learning can have a substantial impact on student achievement. When students in the Long Beach Unified School District—a diverse, urban district in California—used our Khan Academy mastery learning technology for more than thirty minutes per week, their outcomes were associated with a twenty-two-point gain on the state math assessment—a result that was twice the district average. I'm particularly heartened that the gains hold true regardless of race, ethnicity, gender, family income, or English-language learner status. Helping students who are underserved is core to Khan Academy's mission. Bolstered by the findings, we're working with other urban school districts in places like Houston, Detroit, Louisville, and Compton, as well as those in smaller cities, to help more students learn more deeply.

Mastery learning is gaining a foothold in communities across the United States, and there's strong evidence that it can help children around the world. It may, in fact, be critical for students who are living in poverty in developing nations. Two Nobel Prize–winning economists from MIT have shown that filling learning gaps is one of the few educational interventions that works for impoverished children. They told the BBC, "What's really critical is that the kids should have some time when they can catch up with the material they have missed—something that is excluded from most school systems in the developing world."

If we can equip more schools and more teachers with mastery learning technology, imagine the possibilities for students living in poverty in India or attending a large urban high school in California. After closing their knowledge gaps, they can build on their stronger foundations to reach new academic heights. Year after year, they'll make additional progress and break new educational ground. Mastery learning isn't a silver bullet for all students, but I deeply believe that it's a powerful approach to learning that's ready to be unleashed.

It also has huge potential for adult learners, who might be embarrassed to go back and learn stuff they should have mastered in high school. Now they can spend a few hours a day getting up to speed online and stop feeling sheepish about what they may or may not know. They can self-educate, reeducate, or retrain themselves. They can tap into their potential to become adaptable citizens of a new world, equipped with a mastery of math, writing, science, or computing.

Ultimately, our goal is to use technology to humanize what's happening in education on a global scale. Much like the Industrial Revolution, there is potential for most people to participate in the current technology revolution, with vast implications for the progress of civilization. Education will allow people to move into new professions with mastery and confidence. Millions of Americans will be empowered with the skills they need to participate in what I believe will be one of humankind's most exciting centuries. That will be a history lesson for the ages.

Making Mischief

CARISSA CARTER AND SCOTT DOORLEY WITH DAVID KELLEY

Carissa Carter is a designer, educator, and writer and the director of Teaching + Learning at Stanford University's Hasso Plattner Institute of Design, known as the d.school. Scott Doorley is an author, designer, teacher, and speaker and the creative director at the Stanford d.school. David Kelley founded the global design and innovation company IDEO and the Stanford d.school.

Put seven folds into a piece of paper in just the right places and you've made something that can fly. You are part of the collective of makers. All because paper is a material that makes sense. Fold it, rip it, cut it, twist it, whatever—you can pick it up and feel how it works.

As designers, we make things with materials. Traditionally, these materials have included metal, wood, plastic, pixels, storyboards, spaces, and beyond. We make products, experiences, systems, places, programs, movements, and companies. We're physical and digital. Boutique and mass-produced. Design, today, is a broad discipline. And the materials of making are more complex, interconnected, and unpredictable than ever.

Today's materials of design include tricky technologies. They are algorithms and blockchains, synthesized organisms and DNA sequences, massive data sets and social networks. These materials are opaque and unruly. Most of us don't look at an app and realize what algorithms make it work. Most of us can't tell that the ear of corn on the left has a few bonus genes spliced in from the one on the right. We can't pick up a blockchain and figure out how to fit it into a social network. Today's materials make it extremely difficult to grasp how the products, experiences, and systems around us work—or even recognize whether they're working at all.

We're in an era with a new class of technological media for makers. We call these mischievous materials.

Mischievous Materials Are Hard to Read

Material literacy is critical.

Consider this:

With a stack of reeds and a bit of guidance, you could probably weave a decent roof. With a pile of stones and some trial and error, you could make a workable wall. With a little effort and exploration, you could envision how to build a house out of wood. With the right wood, a few simple rules for spacing studs, some limits for the lengths of joists, and some triangular bracing, you can learn to think in post and beam. These materials make sense. If your house tips over, you know. The materials are transparent and hand-manageable.

Not so with a pile of algorithms. Here's a big data set full of photos of different neighborhoods around the world. Now, go figure out how to build a house. Go ahead and tinker your way to the right combination of dimensionality reduction, regression, and reinforcement-learning algorithms to get the desired outcome. But how do you know what an algorithm does? How do you play with it? How do you figure out its properties, its potential? How do you roll it around in your fingers? How do you know it's not broken?

Do you even know where to begin? If you're building a house or planning a neighborhood, it makes sense that you'd want to know all of the materials you could possibly use. Yes, some might be too expensive or not ideal given the landscape and local climate. But if you were building a home and didn't know that wood was an option, doesn't that feel like a miss? If you were building a house and didn't know that you could plan to put windows in certain places to maximize light and minimize electric costs, doesn't that feel like a miss? If you don't know what the materials are, you can't contribute to the discussion. If you don't have a sense of the types of outcomes they might bring about, you can't envision all of the possibilities. Of course, a homeowner need not be an architect or coder, but the best homes will be built by people who are at least literate in all the possible materials. Anyone who's not at least conversational in the language of a material can't speak and won't be heard. Their ideas and experiences won't be represented and can't be created if they can't imagine with the material in mind.

And when (not if) something goes wrong, those who are material-literate will recognize that and know where to go for help. Pipe bursts, see a pool of water, call the plumber. But what does it look like when your smart home starts selling intimate details about your personal life to foreign governments? Who do you call?

Mischievous materials are hard to read, and it takes time and investment to learn how. In design and technology, when we talk about scale, we are usually referring to scaling consumption or adoption—getting more people to use our products and services. That's worked all right for the last century, but scaling capability and material literacy, especially with mischievous materials, is what will ultimately allow all of us to thrive in the future.

Mischievous Materials Scale Fast and Loose

We need to stop talking about shipping products and start talking about shepherding offspring. Mischievous materials scale rapidly and unpredictably, and our processes for handling the products (offspring!) they create are dated.

Most companies don't make static products anymore. They haven't for a while. The things they make now are much more fluid—even a "fixed" product is actually almost always a string of evolutions. In software, these are called versions or builds, and they usually have wonky numbers like 10.3.2. Yet even in software, we call delivering a product "shipping." Of course, no actual shipping is involved. Yet we still think of "shipping" to "users" or "consumers."

All these terms are outdated. "Shipping": In a digitally networked world, nothing "ships." Turn on auto-update, and the product changes overnight, on its own. "User": Nowadays, a user is as often a creator as a consumer. "Consumer": This term should have been thrown in the rubbish bin (or preferably the compost heap) long ago. Our habit of consumption is slowly, but finally, going out of style.

Those rolling releases and constant updates in the digital world are examples of shepherding. The nagging issue is that they're only designed to shepherd in a few dimensions: user experience, efficiency, and profitability.

Of course, we are quickly coming to understand a side effect: Things can go haywire at scale. Remember when social media was connecting the world and bringing us together? We forget, but in many ways, it did and does. Now, it also seems to be breeding intense social anxiety for a generation of youth and proliferating enough disinformation to test the strength of our social fabric, if not tear it! At scale, digital media behaves more like the weather—hurricanes, floods, and all—than like a factory: complicated, but still predictable.

Our new, mischievous materials—artificial intelligence, biology, and distributed public ledgers—continue to morph after they are made or molded. We've entered an era when it's truly impossible to predict the impact of our creations because they change us as fast as we make them. Put something new in the world, and it will ripple far beyond the neighborhood pond. It will grow, heal, evolve, and maybe even destroy. Possibilities will be magnified. Mistakes will be amplified. Either way, this runaway design is our future. With intentional creation, we may thrive. Without it, we will certainly suffer. Gulp.

Because they change, because they interact in systems, because they have untold complexity, mischievous materials act more like "nature" than like objects and products.

It's time to start treating them that way by building healing responses into our designs from the get-go.

We already build lots of things that need to perform in unexpected circumstances. Once again, take a house. A house is one of the most inflexible and intimate creations we build. Primarily, a home is a shelter. It might also be designed around the stylistic whims of the homeowner. It may be a McMansion-style status symbol, a neoclassical nostalgia trip, or a Scandinavian modernist functional flat. Homes are also designed to withstand. If it's a house in California, it can survive most earthquakes. If it's in Miami, its windows should come with storm shutters. It always has gutters and drainage. Despite all these fixtures, its design also allows for it to be repaired and remodeled over time.

We already encourage building the response into the design through things like building codes and best practices. When we design homes that interact with the unpredictability of the weather, we design with the response in mind. Why not with new materials?

We could—and should—be even more aggressive about building the response into our designs from the get-go. We can build self-healing mechanisms into all design from here on out. If a material fails, we don't just rebuild, we respond. Self-healing can be built in through policy and by design. With nature again as our example, imagine we institute self-healing practices into our response to climate events. With each hurricane, with each flood, rebuilding occurs with local labor and craftspeople (to heal the economy), using sustainable practices, materials, energy systems, and algorithms, blockchains, and beyond (to help heal the underlying problem). Little by little, we heal the material world the same way our bodies heal our broken bones.

What is the algorithmic version of building a healing response into the design? Let's figure it out. If spreading known falsehoods in the digital social media world has real-world consequences, not responding is depressingly dim-witted.

We could algorithmically compost ad dollars. When a bit of media is shown to be an intentionally spread, known falsehood, its earned ad revenue gets redistributed to sources that have a historically high rate of veracity. Compost the lies to fertilize the truth. This reuse would do double duty—encourage fact-seeking, discourage lie-peddling. This solution feels immediately dicey even as we write it down. We can hear the arguments forming in your head. Wouldn't that spell the end of free speech? Who gets to decide the truth? Nobody knows yet. Let's do a little digging and see what works. Ironically, as we move into the digital world, the only way to respond is to drop our binary all-or-nothing thinking, embrace the reality of our mischievous materials that scale fast and loose, and dabble in new ways to till our digital soil. If you want to work in the garden, you've got to be willing to get your hands a little dirty.

Working with Mischievous Materials

There is one technology that we haven't touched on, but it is the elephant in the room: us. *We* are the technology we should be most concerned about. The rate of change has finally surpassed our ability to understand what is really afoot. We're ill-equipped for the next twist of fate. For better or worse, we're also the operating system that everything else runs on.

We're in a unique moment, but this must be what it felt like to be alive during the dawn of the Industrial Revolution. Progress, in the form of technology, was changing everything. What once was valuable—certain types of labor, certain ways of communicating—was becoming obsolete as the mechanical world took over our hand-hewn history. New noise, new pollution, and new possibilities were usurping old ideals and idyllic pastures. On the surface, the tangible technology kept changing, but the real change was intangible and more profound. Those early machines changed the trajectory of our planet and our species.

This potential for change is why we love design. It's also why we should be humbled by it. Designers are not inventors; we are tinkerers, assemblers, remixers. We work with materials, but what we're really doing is influencing actions, feelings, and culture. We create in order to rearrange invisible things—relationships and behaviors. In a way, the materials are . . . immaterial. They seem like the end product, but they are always just the medium. The Latin root of *medium*, "medius," simply means middle. The materials are in the middle—between us and what we're actually doing. That little distinction is worth some attention.

The history of design (and technology) is a romance. It's a story about the things we make and about how those things make us. Like any real love story, it has its ups and downs: its giddy young love and late-stage heartbreak. In the end, it's still a relationship.

In our love affair with our mischievous materials, we're mistaking the medium for the purpose. We've confused the tool for the result, but the means is never the goal. Social media has never been and never should be the purpose. "Making the world more connected" is not the purpose, either. That's just an affordance, one thing the medium can do. A beam can support some weight.

That's great, and "supporting the weight of the world" would make a nice tagline for a lumber company. But the purpose in using the beam is not supporting weight; it's creating shelter for a family. Social media can connect people. So what? Our goals should be something else. These goals are precisely where we're falling flat on our faces.

We're dazzled by our own creations, but we're mistaking the body for the soul. We're kids riding a bike for the first time screaming, "Look what I can do!" That's fun and functional, but it's step one. It's not the real purpose. Function is a question of can. Purpose is a question of should.

We love the immediate outcome, and we're great at instant gratification. It's not enough. Our goals and their repercussions are bigger than we let on. When we think we're doing one thing, we're actually setting many things in motion. We really have little choice but to build not only with our desired immediate outcomes in mind, but also with the potential second- and third-order effects of our work, positive and negative, in mind, too.

The first step—as we continue our relationship with these new, mischievous materials and their mercurial offspring—is admitting how little we know about what is to come, how ignorant we actually are. Instead of reacting with cliché utopianism, as we did a few years back, or knee-jerk fatalism, as we're likely to do for a few years to come, we might be better served by responding with a bit more humility. Accept that we don't know and test some things out.

With that humility intact, here are a few ways we might move ahead:

Design in a diverse, rich environment. Get as many people as literate in the language of design and the lexicon of mischievous materials as possible so we have a diversity of perspectives designing for a diversity of needs. Design not just "for" the people but "by" the people.

Start to relate to these new, unwieldy materials more like they are part of our natural world—cultivate, nurture, and prune them. Think less of shipping end products and more of shepherding them along their life cycle.

Last, with a plain-sighted recognition that these things will unfold imperfectly, we can evolve and adapt them as they come into being. We can build healing mechanisms into their design and into our habits the way wine growers plant mustard during the off-season to renew the soil and protect against pests.

New, tricky technologies are today's design media. They aren't ends in themselves. Their future, our future, is (and will ever be) only partly in our control. Let's get better at understanding mischievous materials—the what, how, why; the injustices; the injuries; and the infinite possibilities. And when we use the wrong material, when it slices us open, we'll know how to heal our way forward.

Has Coronavirus Made the Internet Better?

JENNA WORTHAM

Jenna Wortham is a culture writer for the New York Times Magazine *and cohosts the* New York Times *podcast* Still Processing.

On a Saturday night in 2020 while COVID-19 was in full swing, Derrick Jones, a DJ who performs under the name D-Nice, livestreamed himself working his turntables from his home in Los Angeles, where he was self-isolating. He started early in the afternoon and played deep into the night, pausing only to sip his drink, take the briefest of bathroom breaks, and change into a new flamboyant hat. Despite all the chaos outside, here, online, was a safe harbor. The only thing contagious was the mood, which was jubilant. As names of friends–and increasingly, famous people–floated across his screen, he would grin and call out their names in greeting: Rihanna. Dwyane Wade. Michelle Obama. Janet Jackson. As the night stretched on, the party shifted into something more meaningful than a celebratory distraction. Time and space collapsed as tens of thousands of people experienced the same song, the same shared spirit, no matter who or where they were. Kind of like COVID-19 itself.

At one point, apparently inspired, Jones shouted out thanks to all the nurses, doctors, and hospital workers. His eyes drifted to the number of people in the "room," surging toward 150,000, and paused, amazement shaping the contours of his eyes and mouth. "We should raise some money or something," he said.

What D-Nice seemed to realize in that moment was something many people have realized since COVID-19 gripped the country: Social media could be mobilized for something far greater than self-promotion. Artists have taken to YouTube or Instagram to provide some relief, to allow us to gather together and listen to an opera, or hear a standup set, or watch a poetry reading, all of us separate but still together. But more remarkable, it has become the medium by which people have organized to help others.

On Twitter, writers like Shea Serrano and Roxane Gay helped raise money for bills and groceries for those who are struggling. Programmers connected online to create a tool to schedule cooperative child care. Prison-reform organizations worked to bail out incarcerated people and send hand sanitizer to prisons and jails, where the virus is rampant. Google Docs files began circulating with information on food pantries and how to apply for unemployment. GoFundMes quickly popped up to distribute money to people hit hardest by the crisis, including sex workers, restaurant workers, and underinsured artists. Healing practitioners made meditation sessions, yoga classes, and other mental-health assistance available free online. Sewing patterns for masks and surgical caps were circulated online, and everyone from the rapper Future to the designer Collina Strada began efforts to produce them for frontline workers. Copper3D released its pending patent for 3D printed masks, allowing anyone with a printer to churn them out and distribute them. In my own neighborhood, someone created a Slack channel where people shared strategies for deferring credit-card payments and rent and offered to run errands for families in need. Even the online performances, like D-Nice's dance party, felt as though they were really less about pure entertainment and more about serving a nation, a world even, that was suffering in isolation and fear.

For a time, futurists dreamed, optimistically, that cyberspace might exist as a place where humankind could hit reset on society. The idea was that the arrival of networked computers would create an imaginary space where bodily markers of difference would be masked by a utopian fog. In 1996, at the World Economic Forum in Davos, Switzerland, John Perry Barlow issued a manifesto titled "A Declaration of the Independence of Cyberspace," which stated, "We are creating a world that all may enter without privilege or prejudice accorded by race, economic power, military force or station of birth." Barlow continued that the civilization he and others hoped to create would "be more humane and fair than the world your governments have made before."

By now we know that those dreams were a fantasy, informed by the same imperialistic and colonial urges that underpinned the creation of the internet itself. No dream internet utopia ever emerged. Instead, societal woes have been compounded by the rise of technology. The internet has been oriented around an axis of maximizing profits, almost since its inception. In *The Know-It-Alls,* the journalist (and my former colleague) Noam Cohen documents the emergence of Stanford University (nicknamed "Get Rich U.") as the birthplace of Silicon Valley, a place where a "hacker's arrogance and an entrepreneur's greed has turned a collective enterprise like the web into something proprietary, where our online profiles, our online relationships, our online posts and web pages and photographs are routinely exploited for business reasons." Today, it feels almost impossible to imagine another way of thinking about the internet.

And yet, in the aftermath of the arrival of the novel coronavirus, one has emerged that feels, at least for the moment, closer to John Perry Barlow's embarrassingly earnest speech. It's worth noting that he also said that cyberspace was an "act of nature, and it grows itself through our collective actions."

Historically speaking, new infrastructures tend to emerge as a response to disasters and the negligence of governments in their wake. In the 1970s, for example, an activist group called the Young Lords seized an x-ray truck that was administering tuberculosis tests in East Harlem, where the disease was prevalent, and extended the operating hours to make it more readily available to working residents. In the days since the crisis began, I've been turning to Adrienne Maree Brown's 2017 book *Emergent Strategy*, which offers strategies for reimagining ways to organize powerful movements for social justice and mutual aid with a humanist, collective, anticapitalist framework. She describes the concept as "how we intentionally change in ways that grow our capacity to embody the just and liberated worlds we long for." Her book asks us not to resist change. That would be as futile as resisting the deeply embedded influence technology has on our lives. It's the same as resisting ourselves. But rather, it asks that we adapt, in real time, taking what we know and understand and applying it toward the future that we want. The internet will never exist without complications—already, many of the tools that are helping acclimate to this new cyberreality have been called out for surveillance—but perhaps people are learning how to work the tools to their advantage now.

A few days after his marathon set, Jones talked to Oprah (over video) about his experience. "I've been in the music industry for over thirty years . . . but nothing felt like that, helping people." Shortly afterward, he announced that his next party would be a party with a cause: a voter-registration drive. In one night, he helped motivate thirteen thousand people to start registering.

Making Internet Platforms Accountable

ROGER McNAMEE

Roger McNamee is the author of the New York Times *bestseller* Zucked: Waking Up to the Facebook Catastrophe, *which documents his journey from mentor to Mark Zuckerberg to critic of Facebook.*

Paralyzed by polarization, policymakers in the United States struggle with the most basic functions of democratic government. The country cannot agree on facts, much less policy related to existential issues like climate change. Policymakers and voters are beginning to understand that internet platforms exacerbate polarization while undermining elections, public health, privacy, and competition in the economy.

The convenience of always-on connectivity acts like a narcotic for users and advertisers, creating an illusion of indispensability that has led to near-universal adoption. Blind trust in technology left users vulnerable, and bad actors took advantage of them. Fortunately, journalists and academics have risen to the occasion, and none more so than Harvard professor Shoshana Zuboff, whose book *The Age of Surveillance Capitalism* revealed the economic model underlying the largest internet platforms—naming its key elements and framing the challenges they create. Two years of near-daily revelations dispelled the myth that the products of Google, Facebook, Amazon, and Microsoft are benign. Their business model, algorithms, and culture have aided in the undermining of democracy, public health, privacy, and competition, forcing users and policymakers to confront unpleasant truths. In less than three years since the first stories and investigations broke, awareness of the dark side of internet platforms has reached critical mass. The flood of disclosures since October 2018, when I completed the manuscript for my book *Zucked: Waking Up to the Facebook Catastrophe*, has transformed our understanding of the threats they pose.

The complexity of those threats derives from the platforms' ubiquity and convenience. Success created scale and wealth, giving them economic and political power unrivaled in the economy. Internet platforms may have contributed to multiple instances of election interference; a "textbook ethnic cleansing"; at least one act of terror; numerous mass killings; outbreaks of infectious diseases that had once been eliminated; increases in harassment, bullying, and potentially teen suicide; and many other harms. Few policymakers and users can see a path out of the current mess. Even if policymakers could get past their reluctance to take on complicated issues, few know where to start.

Policymakers face challenges related to the design, business practices, data strategy, and success of internet platforms. The platforms create at least four classes of harm—to democracy, public health, privacy, and competition—that affect individuals and society as a whole. The harm goes beyond consumers to affect all of society. Data is replacing oil as the commodity at the heart of the economy, an opportunity created and dominated by internet platforms. Major industries are surrendering their future to these platforms, apparently unaware that the fate of newspapers may await them. Governments, strapped for cash, are turning over public services to internet platforms without understanding the implications of surveillance capitalism for democratic processes and self-determination. Doing nothing about this would be catastrophic, and doing too little would be almost as bad.

Consistent with today's policy norms, prescriptions have focused on individual symptoms. For example, the EU's General Data Protection Regulation (GDPR) and the California Consumer Privacy Act (CCPA) address a subset of the privacy issues associated with internet platforms. Attempts to change undesirable behavior with fines have been ineffective. Investors, for example, greeted the Federal Trade Commission's record $5 billion fine against Facebook—for privacy violations related to the Cambridge Analytica scandal—as good news, thanks to the settlement's blanket immunity to the company and its executives for other privacy violations. The European Union's two multibillion-dollar fines against Google for antitrust violations also proved to be ineffective. From these examples, we can deduce that the benefits to platforms of violating norms and rules related to anticompetitive behavior and privacy greatly exceed the costs that governments are able to impose on them.

The situation in the developing world is even worse. Myanmar struggled to get Facebook to take action after its platform was exploited to enable ethnic cleansing of the Rohingya minority. Sri Lanka had a similar experience after hate speech on Facebook triggered violence. In the end, Sri Lanka shut down internet platforms temporarily to get their attention.

The evidence to date supports the view that platforms are more powerful than governments. Until the government of a major economy is willing to follow Sri Lanka's lead, the platforms will act with impunity. They may be correct in their belief that users will not allow regulation for fear of losing convenient services. It is up to governments—the US government in particular—to prove them wrong.

The fix for internet platforms begins with recognizing that consumers and society have suffered great harm; that platforms have ignored many opportunities to reform themselves; that the economy would benefit from reducing their monopoly power; and that past federal intervention in the technology industry has produced better outcomes than is generally understood. As with climate change, reform of internet platforms would create a huge economic opportunity, a Next Big Thing. Since the first antitrust action against AT&T in 1956—which created the independent computer industry and launched Silicon Valley with the free licensing of the transistor—government intervention has proven to be a huge stimulant to innovation and entrepreneurship. In the 1960s, the IBM antitrust case fueled the software and personal computer industry. Follow-on cases against AT&T created the data equipment, broadband data, long distance, and internet industries, and the Microsoft case effectively enabled Google. Contrary to conventional wisdom, regulation has helped startups and promoted innovation.

Policymakers will have to pass new legislation to reform business models and use antitrust intervention to undermine monopoly power. Those changes will unshackle entrepreneurs, stimulate innovation, and set in motion the next big market opportunity in technology. This is a rare case where policymakers can be bold, with confidence that what they are doing will improve the economy in the medium and long term.

Shoshana Zuboff exposed how Google and others have leveraged massive data collection, machine learning, and artificial intelligence to gain dominance in the economy. Every action we take these days—whether it be in banking, health care, transportation, or on the internet—leaves a digital footprint behind, many of which are for sale. Spaces that had once been sanctuaries, such as homes and public areas, are increasingly saturated with surveillance technology. As reported in the *New York Times*, today's smartphone technology has enabled the tracking of anyone, anywhere, in the physical or virtual world. Surveillance capitalists process all of this data into data models—or, as former Google data ethicist Tristan Harris calls them, data voodoo dolls—that can be used to predict behavior. Those predictions have transformed marketing, providing insights about the timing of demand that has never previously been available. In addition, internet platforms use data voodoo dolls to refine search results and news feeds, steering users toward desired outcomes. They have also experimented with behavioral manipulation of individuals and large populations. Left unchecked, the internet platforms will be able to extract economic value from industries they target, just as the trusts did in the early twentieth century, prior to antitrust enforcement.

Many consumers do not understand why activists like me worry so much about data. They like the services they get from internet platforms and do not perceive much downside. Some say, "I am a digital native and don't mind trading my data for useful services; besides, my data is out there, so it's too late." There are several flaws in this reasoning. First, the data that users provide in the context of interacting with an internet platform is just a tiny percentage—perhaps less than 1 percent—of all the data that platforms have about them. Second, platforms are using data voodoo dolls to manipulate consumer choices in ways that escape notice, including search results. In some cases, they are able to manipulate behavior. Third, the damage from surveillance capitalism falls disproportionately on the innocent. The victims in El Paso, Pittsburgh, Christchurch, and Myanmar did not need to be users of internet platforms to be murdered; their fate was determined when killers engaged with extremist content. To the extent that manipulation of internet platforms influences elections, the victim count jumps into the millions. Fourth, the internet platforms are developing new businesses that leverage surveillance capitalism, in areas such as smart cities, cryptocurrency, collaborations with law enforcement, and health care. If successful, each of these projects will reduce the ability of consumers to control their own choices and of governments to protect the public interest.

It is true that personal data is "out there." There are few restrictions on the gathering and exploiting of personal data for commercial purposes. That said, the status quo is not inevitable. There was a time when the consequences of irresponsible handling of chemicals caused massive harm to the environment and public health. Changes in policy forced new business practices and the cleanup of toxic spills. I propose that the same notion be extended to personal data. Corporations should no longer be able to collect, retain, trade, and monetize personal data outside of first-party, intended-use transactions. It should be illegal to use data voodoo dolls.

Algorithmic amplification is one of the root causes of harms caused by internet platforms. Platforms argue that it is necessary to "give users what they want." That might be a reasonable argument if consumers actually wanted hate speech, disinformation, and conspiracy theories. The vast majority do not like such content, but they cannot help but react, since it triggers psychological processes associated with fight or flight. While algorithmic amplification may promote helpful content some of the time, the evidence suggests that is the minority case. Particularly disturbing evidence comes from the YouTube recommendation engine, which probably promoted Alex Jones more than the combined reach of America's largest journalistic outlets.

Microtargeting is another root cause. It is the key to monetization of internet platforms. The supply chain begins with surveillance, which enables data voodoo dolls, which in turn enable behavioral predictions and manipulation. Platforms capture signals of major life events—a car purchase, marriage, birth of a child—from hundreds of millions of consumers and analyze the pre-event data to identify common patterns. When any user follows a similar pattern, the platform can determine a probability of that life event. A 50 percent probability of a car purchase is worth far less than an 80 percent probability, but even middling probabilities are worth more than traditional advertising. Behavioral predictions can be creepy, however. For example, researchers discovered that internet platforms sometimes know that a woman is pregnant before she does. They can sell that prediction to anyone, including, possibly, to those who promote the antivax conspiracy theory. High-confidence predictions of consumer behavior revolutionized consumer marketing, which, combined with microtargeting, explains the exceptional growth and profitability of internet platforms.

Two failure modes of microtargeting have been particularly problematic. First, in the context of democracy, microtargeting has enabled bad actors to spread hate speech, disinformation, and conspiracy theories to increase polarization and suppress voting. Microtargeting of such content is hard to spot because it reaches only a carefully selected audience. This is why activists and some policymakers expressed outrage at Facebook's 2019 policy change to eliminate fact-checking in ads by politicians. Facebook's attempt to hide behind the First Amendment is disingenuous for the same reason. The second issue with microtargeting is that it distorts the market economy. One of the core tenets of capitalism is that there must be uncertainty on both sides in order for a market to function. When one side has perfect information and the other does not, market failures can occur. Internet platforms magnify the asymmetry because they are the primary sources of information for users. Worse still, platforms use data voodoo dolls to massage search results and news feeds in ways that increase the value of the behavioral predictions they sell. Platform customers—marketers—have perfect information, while users have only the information that platforms choose to give them. Users mistakenly trust search results and news feed content, believing the platforms to be honest brokers. That is not the case.

Microtargeting can lead to positive experiences for consumers, as with ads for products that are desirable, but the harms outweigh the benefits. The worst outcomes with microtargeting occur in politics, as well as fraud, sexual predation, violent extremism, and other forms of crime. The challenge for policymakers is to find a balance. In the context of democratic elections, microtargeting has demonstrated huge value for fund-raising and messaging, but also voter suppression. It played a key role in the Brexit referendum in the United Kingdom and the 2016 presidential election in the United States. Shortly after the US election, a Facebook executive bragged to a former colleague that the Trump campaign enjoyed a reach per dollar advantage of sixteen to one over Clinton. I hypothesize that Trump's team gained this advantage by microtargeting highly inflammatory messages, which were then shared and amplified algorithmically. (Once an ad gets shared, Facebook treats it as "organic" content, so it is no longer flagged as an ad.) Facebook employees worked side by side with employees from Cambridge Analytica, Google, and Microsoft inside Trump's digital headquarters—presumably using the Cambridge Analytica data set of thirty million American voters whose Facebook IDs had been paired with voter files to create the most detailed custom audience in the history of politics. (The Clinton campaign chose not to work with platforms in this way.)

Microtargeted advertising has played a key role in the recruitment of terrorists, the spread of the antivax conspiracy theory, climate-change denial, and a host of other issues. In combination with algorithmic amplification, microtargeting gives small numbers of bad actors disproportionate influence in society. The damage experienced to date justifies aggressive policy changes addressing root causes, irrespective of the consequences for the profitability of internet platforms. Such policy changes are a necessary first step toward rebuilding democracy.

As policymakers and voters consider the appropriate role of data in society, I would recommend one guiding principal: Personal data should be a human right, not an asset to be monetized. "Own your data" and "data portability" are flawed concepts that address symptoms. They would not repair the damage to institutions and society from surveillance capitalism. In practice, accepting data as a human right would dismantle today's data economy and force a restart with different values. It would not eliminate the use of data, but it would restrict it severely—at least at the beginning. For example, there would be no restriction on a consumer's ability to give a ride-sharing firm his or her identity, billing information, location, and destination to procure transportation, but the corporation would not be able to pass along any of that data to a third party or use it in any other context. I call this "first party, intended use" of data.

Zuboff compares this issue to the debate about child labor in the early twentieth century. In the nineteenth century, children as young as six worked up to thirteen hours a day. The early debate about limiting child labor focused on the number of hours children could work in a day, just as the current debate about the use of personal data focuses on relatively small changes to existing practice. Eventually the conversation about child labor changed, recognizing that any form of child labor was inherently exploitative. Zuboff makes the case that any economic exploitation of personal data by third parties is no different than child labor. It should not be allowed.

Humanity's talent for developing new technology has far outstripped the ability of users and society to adapt. This is true of internet platforms and artificial intelligence. Blind trust in new technology has undermined democracy, public health, privacy, and competition. Children and adults are not equipped to protect themselves. Systems of democracy are failing under pressure from the products in the market today, with more direct assaults on the horizon from new technologies. The leaders of major platform companies have not demonstrated any awareness of the responsibilities that come with the great power they wield. If the country is to have any hope of repairing democracy, and if we hope to address existential crises like climate change, policymakers should prioritize the reform of internet platforms. A new phase is beginning. If we address the flaws of internet platforms and surveillance capitalism, we will be taking the first steps to restoring the more perfect union envisioned by the nation's founders.

Restructuring the Tech Economy

JARON LANIER

Considered a founding father of virtual reality, Jaron Lanier is a computer scientist, visual artist, composer of classical music, and author of Ten Arguments for Deleting Your Social Media Accounts Right Now.

There are approximately four ideas for how society can improve its tech giants: privacy rights; antitrust enforcement; a new kind of tax on "tech," since tech is centralizing a new kind of power and wealth; and a new way for individuals to participate in a tech economy that would spread out the wealth to all the people who created that economy, such as by paying people for their data.

I've come to favor the fourth option because the first three don't directly address the worst problems.

People from all over the political spectrum are furious with the current batch of social media companies, which I'll define to include Google and its subsidiary YouTube. The primary discomfort generally reported is not with economic power or data insecurity, though those issues also trouble a great many people. Instead, there's a primary sense of the world being darkened, of the worst corners of human nature being amplified at the expense of our sanity and survival.

The 2019 massacre in New Zealand, for example, was designed for Facebook streaming by an individual who was radicalized on social media. High-tech darkness amplification is not partisan, but universal.

Tech platforms uniquely influence who gets to say what to whom and how loudly. Since the prime directive for such a company is to accelerate "engagement," the most piercing and annoying speech will tend to be the most favored. Thus, the most irritating people, those who trigger the deep emotions related to the primal behaviors of fight or flight, tend to be amplified the most. The nastiest side of human nature becomes dominant.

Nice people are amplified, too, of course, but they tend to become the fuel for a higher volume nasty backlash. Black Lives Matter, which I supported, added data to the system that was used automatically and perversely to identify, introduce, and spur on a renewed KKK and neo-Nazi movement in the United States, for instance. That was engagement amplification in practice.

There are endless jeremiads about how the current nature of networking is bringing out the worst in people, but these usually end with a tragic sense of resignation. The problem is the core business model of companies like Facebook and Google. It's too late, runs the familiar conclusion; we're stuck from now on with our worst selves forever.

But even that isn't the worst of it. The same companies proudly assert that they are in a race to dominate artificial intelligence. All the data gathered from the people who are being made nasty, through the process of engagement, is being used to train the AIs that will put those same people out of work. In the meantime, the soon-to-be-obsolete humans can get temp work at tech companies like Uber, which seem optimized to get rid of the people as soon as AIs can take over.

The crisis therefore goes beyond emotions and omni-defamation to core spiritual identity. At a recent gathering of high school students, I was asked the darkest question I've ever heard from a teen: "If AI is going to take the jobs, why did our parents have us? Why are we here?"

Fanatics like the New Zealand shooter cling to blood and soil; they perceive no other option for finding validity. Everything else has turned into a meme.

It is impossible to direct a social media company to block content that inspires existential dread. There is no way to define that well enough for an algorithm, a human moderator, or a court of law.

Perhaps a change in the business model could help. If data is the new oil, and data comes from people, why not pay people for their data? Free online video is more often sadistic than paid video. Maybe paying and being paid can be an avenue for improving civilization.

What if people were owed money for the use of their data when they were targeted by a political ad, enough that such targeting was no longer a viable business model? What if people could earn money from contributing data to AI? Might they not develop a justified sense of pride in helping to program the robots? Might the data and the robots not perform better once people are awakened to their new roles? Might not the economy expand greatly once we admit that a lot of people are productive in new ways? What if people find it easier to find meaning in a world that tangibly values them?

If the business model of companies like Facebook is the core problem, then surely the way out of our mess must be to change the business model. Currently, the model is that all human activity on networks is financed by third parties who hope to influence the immediate users. How can that model lead to anything other than a world optimized for manipulation and unreality?

Just as users pay for Netflix, engendering a new era of "peak TV," they'd start to pay for a new era of peak social networking that doesn't amplify darkness. Those who cannot pay would be supported by new public services analogous to public libraries. The same users would accumulate a wide and growing array of royalties from their data.

This is an idea that makes enough sense that certain tech companies have tried to discredit it, even though it isn't yet very prominent in policy circles. Facebook and Google have both stated that data from individuals isn't worth much, but they only say that in special settings. When it comes to arguing for stock value and market caps, then the data race merges with the AI race and is trumpeted as the most valuable aspiration in the world.

We can't expect an online utopia, any more than we can expect an offline utopia. People will always be annoying to one another. But we can and must demand better from tech companies. Tech had until recently been the last bastion of optimism. We could all agree on that one bright spot in our future, but now we only see a cliff we are stampeding toward.

If we want to regain hope, meaning, and even a slight capacity for kindness when we disagree, we must reform our tech world.

Rejecting the Sirens of the "Friction-Free" World

SHERRY TURKLE

Sherry Turkle is the Abby Rockefeller Mauzé Professor of the Social Studies of Science and Technology at the Massachusetts Institute of Technology and the founding director of the MIT Initiative on Technology and Self. She is the author of five books and three edited collections, including four landmark studies on our relationship with digital culture.

More than fifty years ago, during the early days of protests against the Vietnam War, the University of California at Berkeley became famous for its free speech movement. From my perch as a senior in a Brooklyn public high school, I understood this: Students demanded the right to speak their mind about the war or civil rights or university policy without fear of reprisal. I grew up thinking of that campus as sacred space.

Now it's spring 2019, and I'm teaching at Berkeley, the recipient of an honorary lectureship for my research on digital culture. I get to speak and take meals with Berkeley students across all schools. The students I meet are well informed. They are aware of Facebook's complicity with Cambridge Analytica in an assault on privacy. They know that online, untrue things are placed before them and made to look like true things. They understand that in our current information regime, every consumer creates a data stream that is resold for profit. These students say that they have "always" known that Apple, Facebook, and Google try to keep them at their devices the way Las Vegas gaming casinos keep gamblers at slot machines.

Yet I also hear that this is the price of their "luxuries"—the food, books, clothing, films, music, transportation, and communications that their "free" apps summon on demand. *Luxuries*, that's their word. They are critical, from the start, of their use of the word, but I feel a generational responsibility that they use this apolitical word in this political circumstance.

In the late sixties and early seventies, I belonged to a cohort that talked about political power as "the system." We argued that when you made the system transparent and showed how it worked, you could begin to have leverage over it. But we did not step up in the same way when we confronted the technological system. On the contrary, we were smitten and avoided necessary conversations about what it could become.

Indeed, my generation shaped a world where when people looked for solutions, one of the first places they looked was to technology. We made our love of the digital technology that came of age with us central to our identity. The creators of that technology used our generational language of liberation: "Think different." "Do no evil." "Connection as a universal social good." My generation liked being identified with this force for the new.

The crown jewels of the digital revolution—from personal computers and sociable robots to the apps on our phones—shared a vision. Digital technology would change the rules: The difficult will be made easy; the rough will become smooth; that which has friction will become friction-free. Digital technology wasn't just going to make things go more smoothly when you used an app to pay your bills or find your way home. The vision was more ambitious: to minimize and even eliminate social friction, from the face-to-face conversations that are the meat of political organizing to intimate conversations that almost always bring emotional stress. In due course, as we got used to texting rather than talking and to email rather than conversation, "real time" became everyone's enemy, because taking life out of real time meant a life with less vulnerability.

But in that move, technology encouraged us to forget what we knew about life. And we made digital worlds where we could forget what life was teaching us.

Life taught that face-to-face political organizing builds strong connections and institutions. The internet made political expression and organizing exponentially more convenient but just as exponentially more toxic. Face-to-face conversations taught that when we stumble and lose our words, it can be painful, but we reveal ourselves most to one another. Screen life allowed us to edit our thoughts and appear closer to our ideal selves than we knew ourselves to be. We preached authenticity but practiced self-curation.

After a professional lifetime studying where digital technology has taken us, I end up with an intergenerational call-to-arms. For educators. Policymakers. Parents. The young. And the very young. *The value of the friction-free has been overrated. Hiding behind a screen protects you from vulnerability, but a life without vulnerability can be no life at all.* It's time to associate the digital with new values. Instead of smooth and friction-free, the digital can give us greater mastery over the complex, the challenging, and the demanding.

It is time to embrace complexity and challenge in what we fund, in what we write, in what we publish, in what we teach, and in our expectations for relationships.

We can't be afraid to embrace friction. We have to brush up on all those skills we have not been practicing behind our screens. These are emotional skills but also the capability to participate fully in the debates of the public square.

To fix what is broken now we can't look to smoother apps. We need the rough and tumble of the real. So, for our crisis of intimacy and empathy, we need humans, talking to one another; we are the empathy app. When it comes to politics, that, too, is human work, not the work of supersmart machines.

My generation, the generation that came of age in the 1960s and 1970s, took technology as it presented itself—as an opaque force that determined outcomes. Today's students who look at apps and see luxuries are living out the limitations of our vision. Technology, like politics, can be analyzed, made transparent, and brought under control. All of this came out in those conversations at Berkeley, those conversations that balanced current luxuries and those long ago echoes of a campus that went to war over speech.

These days, a call for free speech has to claim the right not only to say what's on your mind but to own it. More than that, when we listen or read on screens, we need to know the provenance of speech. Did someone, a person, actually say or produce it? Is it the product of a simulation? Something a program thought might influence a person just like you?

It took a generation to dispel the notion that digital technology was "just" a tool and that a communication technology that brought "everyone" together could only be a force for good. Now we must face our inventions with new rigor and skepticism. And embrace what we don't share with machines: the human-specific clarity of the friction-filled life.

Unintended Side Effects: Social Media, Walled Gardens, and the Erosion of Democracy

JOHN HENNESSY

Former president of Stanford University, John Hennessy is a computer scientist, academician, businessman, and chair of Alphabet Inc., the parent company of Google.

Most of the early uses of the internet (and its predecessor, the ARPAnet) were focused on communications: email or exchange of files, often of scientific papers or data. The emergence of the World Wide Web changed that, as it enabled almost anyone to execute transactions and be a publisher on the web. Many early users focused on exchanging scientific content, but the marketplace was not far behind. By the time Netscape and Yahoo! were founded (in 1994 and 1995), the internet was making giant strides in electronic commerce. The late 1990s saw the growth and emergence of the net as a global and comprehensive information service. Electronic bulletin boards were perhaps the earliest form of user-generated content sharing; they predated the web, but rapidly grew in use with it. These early systems created interest groups, but not the strong social organizations induced by the later social media platforms.

Friendster, Myspace, and Facebook took the sharing of user-generated content and personal information to new levels and helped accelerate the explosion of the World Wide Web. Today, there are more than two billion websites, more than five billion YouTube videos are viewed every day, and over 25 percent of the world's population is active every month on Facebook. The creators of these systems (as well as the World Wide Web and internet, which underpin them) saw them as enabling many more individuals to publish and share content, but few foresaw that the publication and sharing of information would grow at the rate it has.

Key to this explosion of shared user-generated content was Section 230 of the Communications Decency Act. The rationale behind Section 230 was clear: If internet service providers and web hosts were responsible for the content generated by users, they would have to play a very different role of curation, editing, and even censorship. Section 230 makes it clear that ISPs and web-hosting sites are deliverers, not producers and editors. Without Section 230, it is doubtful that user content could have flourished as it did on the internet.

Walled Gardens and Echo Chambers

While few people predicted the rate of growth in user-content sites such as YouTube, Facebook, and Twitter, even fewer foresaw the unintended consequences. Foremost among these is the creation of so-called walled gardens: communities of users united by some set of beliefs or interests and generally not including individuals with widely differing viewpoints on the underlying beliefs. For example, an interest group promoting gun ownership is unlikely to attract and contain individuals who favor much stricter gun control.

Social media was not the first place that created such closed communities. Cable TV and some publications had already developed user groups that often shared a common set of unifying, non-overlapping beliefs. For example, consider Fox News and MSNBC: They attract distinctly different audiences, few people would watch both, and their commentators are "expected" to express a point of view aligned with the belief system of their viewers. Similar examples can be found in the press, although such publications have largely moved online, using social media–like structures and inducements to engage readers.

While some damage to wide social discourse had already occurred before the social media revolution came along, social media dramatically amplified the effects. In particular, with social media it became possible to create interest groups based on very fine-grained criteria. Although MSNBC and Fox News could build audiences with more liberal or conservative viewpoints, on social media we can build a conservative or liberal interest group that also shares views about abortion, gay rights, or gun ownership.

Interactions within such interest groups are likely to be self-reinforcing—hence the term "echo chambers." Because these groups define their belief systems narrowly, unplanned interactions with people with differing views rarely occur. For example, groups advocating for and against gay rights are unlikely to meet on social media, making it harder to establish shared values on topics that might be seen as divisive. This isolation can lead to more extreme views and a hollowing out of the middle. We have seen this occur on several key issues, such as abortion rights. The majority of Americans support at least limited forms of abortion, while groups on one side are totally opposed and groups on the other side support no limitations. Of course, social media and the internet did not create these divisions, but it is likely that social media has led to the growth of such divisions and the stridency with which they hold their views.

In particular, echo chambers lead to strengthening conformal bias: "I believe X is true. Because I interact with a set of individuals who also endorse the truth of X, my beliefs are confirmed." Rather than hear arguments in favor of more moderate or balanced approaches, my views, as a member of an echo chamber, are reinforced. Moreover, I avoid the friction that my more extreme views might encounter in the broader world, further reinforcing the perceived righteousness of my opinion. By contrast, a colleague and I might discuss an editorial in a major newspaper, agreeing and disagreeing on different aspects. Such serendipitous encounters are much less likely to occur in narrower social media groups—but it is exactly those sorts of broadening encounters that are critical to advancing democracy and ensuring that our country succeeds.

Any possible solutions to the problems of walled gardens and echo chambers confront a strong American bias in favor of freedom of the press (and its modern electronic counterparts). In a strict sense, internet companies are not bound by the First Amendment in the same way that government is bound. In practice, however, there would be a significant outcry if internet and social media companies exercised considerable censorship over content on their sites. Preserving freedom of speech and association are fundamental aims of our society. Indeed, while lawmakers decry the use of social media by foreign governments (in particular) to interfere with our elections, they have strongly advocated for the preservation of diverse views, including more extreme ones.

Any overly interventionist approach, therefore, will generate negative responses. With that in mind, there are more moderate approaches, including restrictive solutions and ones that rely on positive interventions.

Restrictive Solutions

One approach is making internet service providers, web-hosting companies, and social media sites responsible for the truth of the material hosted on their sites. Considering the scale of the network will rapidly convince you that this is an impossible task. How could you verify the truth of more than two billion websites, many of which are continuously updated? In addition, it would require empowering these companies to be "thought police," determining the truth of various sites, when the difficulty is that truth, in this sense, is often in the eye of the beholder. Consider the following statements about climate change:

1. Climate change is a complete fraud.
2. Yes, the climate is changing, but humans are not the primary cause.
3. Climate change is occurring, but the predictions are overstated.

Many people might label the first statement as false; some would certainly label statements two and three as true; and others would consider them all false. Being the arbiter of truth on the internet would be an impossible mission. Furthermore, since many divisive social media sites traffic in opinions, it is unclear that policing for truth, even if it were doable, would be a cure for the walled gardens and echo chambers.

What about a narrower focus: truth in political advertising? The Federal Trade Commission already mandates truth in product advertising, requiring proof for certain types of claims (such as health claims). But if we require political advertising to be true, we could only regulate explicitly false statements. For example, we could prevent statements such as "President Obama was born in Africa," or "Humans never landed on the moon." But we could not remove statements claimed by a small but rational minority, such as, "Global warming is not caused primarily by human-generated CO2 emissions," or "President Trump won the election due to Russian interference." Still, while such restrictions would likely have limited impact, they might be worthwhile, considering the damage of patently false political advertising.

A somewhat more practical solution would be to eliminate the use of microtargeting in political advertising, which can be extremely divisive and increase isolation, reinforce biases, and further undermine broad attempts at dialogue. But restrictions on microtargeting would generate both positive and negative support in both US political parties. Consider that microtargeting could be used both to enhance voter participation (typically by individuals whose voting preferences are known) and to discourage voting by highlighting negative aspects of a candidate that might cause voters to be unenthusiastic about all candidates and not cast a vote. Still, eliminating microtargeting in political advertising has the support of at least one member of the Federal Election Commission.

Positive Solutions

Instead of restrictions, positive solutions aim to find ways to use the internet and social media technologies to promote more discourse and stronger democratic processes. Three possible solutions have been widely discussed.

Virtual town halls create open gardens rather than walled ones, bringing together individuals with different viewpoints and common global interests for electronic town hall–like discussions. In addition to actual town hall meetings, programs such as *Meet the Press* and other TV-based town hall meetings provide some models. Bringing together individuals with disagreeing views before a real or virtual audience can promote a rigorous intellectual debate over the issues, helping to bridge divides and generate understanding. Without significant participation, however, the effectiveness of virtual town halls would be limited.

Other possible approaches include "must carry" rules to increase diversity of programing. In 1965, the Federal Communications Commission passed rules requiring that cable systems must carry local TV channels. Some of these rulings, however, have been found unconstitutional, particularly when a cable network does not cross state lines. Such regulation could prove even more difficult on the internet, where user communities tend to be more global. Assuming those issues are overcome, one could imagine a rule requiring social media companies to host certain sites and programs designed to broaden discussion and participation in the democratic process.

Finally, we could conceive of a system where a federal subsidy was used specifically to create opportunities for broader dialogue among citizens. The challenges here involve ensuring broad community participation and doing so cost-effectively.

In truth, no one really foresaw the unintended and negative side effects of social media on democracy and civic dialogue. While we all thought that liberating anyone to be a content provider would be beneficial, we failed to realize that social media could isolate groups and create echo chambers, exacerbating a problem that cable news had already helped create. The negative consequences of this problem are significant, and solutions will not be easy. While government regulation and, eventually, legislation may play a role in reducing negative consequences, the best hope in the short term will be for the companies in this industry to step forward with solutions. That will take real leadership.

The Change in the Nature of Change

JAMES G. COULTER

James G. Coulter is cofounder of the private equity firm TPG Capital, originally known as the Texas Pacific Group.

As an investor, I have built a career by focusing on the challenge and opportunity of complexity and change. My private equity firm has helped restructure the airline industry, financed health-care solutions, helped clean up the savings-and-loan industry, and participated in the reimagining of the transportation and lodging industry through early investments in Uber and Airbnb. The strategy of "buying" the complexity of change and "selling" clarity once it is understood has served our firm well. We have, out of necessity, built an expertise in pattern recognition and change.

But today's technology-driven business environment has magnified the complexity of change and clouded the search for clarity. Investors have been required to retrain and reorient themselves to a new reality where incumbency is devalued, generational changes happen on intragenerational time lines, and businesses rise and fall with regularity. Business challenges have evolved from planning and analysis to entrepreneurism and an invocation to "move fast and break things." Investors and businesses have had to adapt to confront the change in the very nature of change.

The digital era has ushered in similar levels of unprecedented changes in society: changes in communication, consumer behavior, entertainment consumption, the workplace, our homes, and schools. These are ubiquitous, bone-rattling changes.

This era has not mimicked historical social changes. There is no relevant playbook to consult. The digital revolution that has gathered steam over the last thirty years has transformed the shape and the speed of change and challenged our ability, as a society, to manage and control technology to our ultimate benefit. This "change in the nature of change" has altered all business marketplaces and will likely challenge the marketplace for democracy.

Punctuated Equilibrium

Throughout the twentieth century, our image of change, in business and biology, was rooted in a sense of gradualism. In biology, the constant mutation doctrine of Charles Darwin's 1859 *On the Origin of Species* overcame religious challenges to dominate evolutionary thinking. Similarly, late-twentieth-century business practices were based on long-range strategic plans and five-year capital budgets. Business graduate schools taught *kaizen*, the Japanese method of crowd-sourced continuous improvement. Today, these gradualist models of change and the tools to manage it are almost quaint.

In the 1970s, evolutionary biologists recognized the need for a new model to describe change. Two paleontologists, Stephen Jay Gould and Niles Eldredge, pointed out that Darwin's theory failed to fully explain significant gaps in the fossil record. Their work showed that the story of life on this planet was not the constant gradualism of Darwin, but instead long periods of relative stability punctuated by brief, violent episodes of change. Dinosaurs ruled for millennia and were then gone. *Homo sapiens* arose and changed the long-term stability of the planet in a geological instant. In Gould and Eldredge's words, "The history of evolution is not one of stately unfolding, but a story of homeostatic equilibrium, disturbed only 'rarely' (i.e., rather often in the fullness of time) by rapid and episodic events of speciation."

The "punctuated equilibrium" model—stability interrupted by mutation and rapid species-altering change, which then settles into a new relative stability—radically changed our view of the forces behind speciation. Scientists had to evolve their understanding of the "shape" of evolution. Likewise, the digital revolution has forced us, as investors, to rethink our understanding of the "change in the nature of change" in business. Our twentieth-century training taught us to assume that industries will gradually evolve and improve. In the twenty-first century, business change has gone from gradual to abrupt, from careful diversification to industry-reshaping disruption, and from continuous improvement to discontinuous, uprooted transformation. There has been a distinctive shift in the "shape" of business change, one that echoes the evolutionary implications of "punctuated equilibrium."

The recorded music industry offers a sense of this change in the shape of business change. For decades, the record industry seemed comfortable in an arc of continuous evolution. Genres moved from big band to rock to heavy metal to disco to rap to EDM. Physical delivery moved from vinyl to eight-track to cassette to CD. Still, the continuous "newness" of the music business was anchored in a stable industry structure of record companies, hardware makers, physical media as the units of production, radio-based discovery, and tours as a way of promoting albums.

The emergence of digital delivery upended the nature of change in the industry and shook its very foundations. A decade of uncertainty ensued (iTunes? Pandora? Satellite radio?), along with falling revenues from album and CD sales. Old structures died, and new structures emerged. Today, the industry has reshaped itself into a structure rooted in subscription streaming. Artist revenues are dominated by touring, and albums now promote tours, rather than the other way around. From the chaos of a moment of "punctuation," the industry has emerged into a new growth mode, in a totally unexpected type of structure. A change in the shape of change has entirely altered the music industry and decimated others.

Consider the experience of a taxi driver, a Kodak investor, or a local newspaper reporter in the aftermath of digital disruption of their industries. Like Spanish conquistadors who invaded the Aztec capital of Tenochtitlan (now Mexico City), razed the Aztecs' main sacred temple, and built their own cathedral on top of the ruins, the wave of technological change crashed onto the ride-hailing, photography, and local news industries—effectively wiping them out—and built its new temples (Uber, Apple, and Facebook) atop the ashes.

The orthodoxies of our democratic structures are also under siege by the new shape of change. The long-standing governmental communication protocol built on press conferences, occasional national addresses, and interpretative journalism has been upended by Twitter-based tirades and campaign-like rallies reaching only 20 percent of the electorate, resetting the dialogue of democracy between the government and the governed.

Like business, democracy has to adapt to the change in the shape of change. The traditional tools and orthodoxies of gradualism will fall short of meeting the challenge.

Absence of "Societal Seat Belts"

Society and business have always been shaped by the introduction of new technologies, from the printing press and internal combustion engines to electricity and aviation. The surprise in this chapter of technological evolution is the rapid, almost unbridled global spread of the technology itself. The automobile took seventy-four years to go from approximately 10 percent adoption to 90 percent adoption in US households. The telephone took sixty-six years and the stove fifty-four years. But smartphone adoption by US eighteen- to twenty-nine-year-olds rocketed to 96 percent in less than twelve years.

The more than three billion smartphones parked in the world's pockets are redefining our concept of "speed to market." In the 1450s, Gutenberg spent five years preparing his press to print the original 180 bibles that started a communications revolution. Today, a viral cat video can make it around the world a million times in every direction in an instant. The speed of adoption and change has not merely been redefined—it has been reimagined.

Missing are the traditional guardrails and natural barriers that moderated the speed of change and protected the fabric of democracy. It took more than eighty years after the invention of the incandescent light bulb to achieve its grid-enabled commercialization. Similarly, the air-travel revolution was based on nearly a half century of change, including the evolution of the Wright brothers' plane to Pan Am Clippers, the repurposing of World War II–era DC3s to commercial aircraft, and the creation of a national airport system. While Walmart created a revolution in retailing, it took decades for its stores to spread across the country.

Such physical barriers ensured that generational changes actually played out over a generation. Their slowing effect afforded society and business the time to adapt to and adopt change at a measured pace. Today—unburdened by extensive physical barriers—Amazon grew from $100 billion to more than $200 billion in sales in less than three years; Facebook garnered 2.5 billion users in a little over a decade; and TikTok skyrocketed to 1.5 billion users in a matter of five quarters. With the guardrails of physical expansion removed, the engine of digital change is accelerating unchecked.

Society in general, and democratic society in particular, has always responded to the hazards of technological change, but never rapidly. It took seventy years from the introduction of Ford's Model T before Congress mandated the presence of seat belts. It was fifty-five years from the Wright brothers' first flight before the creation of the Federal Aviation Administration. It took forty years after the first patent medicine company was incorporated in the United States for the federal government to establish the Food and Drug Administration.

We could afford to take our time applying these "societal seat belts" to innovative technologies during the twentieth century because change unfolded at twentieth-century rates. The populace, for example, was less exposed to the hazardous lack of seat belts in the first generations of the Model T because horses still outnumbered automobiles and there were less than five hundred miles of paved road in the United States. Consciously or not, we knew that we had time.

In the twenty-first century, democratic society must react to a technological revolution unfolding on a fundamentally different timescale. How will the inevitable "societal seat belts" be installed, let alone buckled? Are we moving so fast that change will outpace our ability to even contemplate the kind of seat belts that are necessary before they become obsolete? Will we be constantly skating to where the puck *was* instead of where it will be? These are not questions of theory or conjecture.

Out of necessity, investors have quickly learned to adapt to the change in the speed of change. Incumbency has gone from an asset to a risk. The traditional focus on managerial "steady hands" has been replaced with a covetous search for entrepreneurship and disruptive leadership. "How things work today" no longer seems to be a guide for "how things will work tomorrow." Past is not prologue; it is simply an invitation for disruption. The constituencies of democracy will need a similar change in mindset to respond to the change in the speed of change.

Changing How We Respond to Change

The digital revolution has forced investors and businesses to upend their thinking about business models and their approach to disruption. Unquestionably, democracy and society will need to likewise reorient and recommit. Democratic society must fundamentally change its response to change. Otherwise, the technological revolution may cause democratic devolution.

At a time when technology moves, almost literally, at the speed of light, business experience suggests several imperatives for democracy in the digital era:

Face up to the challenge: The last decade has taught business that there is nowhere to hide from digital disruption. It must be responded to quickly and head-on. To lag is, most often, to lose, and democracy has lagged in mobilizing its responses.

The last twenty-five years are woefully littered with attempts at health-care legislation reform, but evidence little engagement in building a legislative response to the societal challenges created by technology. US data privacy policy, for example, has been largely addressed by importing Europe's regulatory frameworks and through state government–driven initiatives. Contrast this with Singapore, which recently announced its Research Innovation Enterprise 2020 Plan, the sixth in a series of five-year plans, dating back to 1995, focused on a "whole-of-nation journey that Singapore is embarking on, enabled by digital technologies."

Before we can even talk about applying "societal seat belts" to these disruptive technologies, we need to build a governmental entity to conceive and administer them. The agricultural industry (1 percent of GDP), the energy industry (6 percent of GDP), and the transportation industry (9 percent of GDP) each has its own cabinet-level departments. We have structures and teams in place to handle the challenges of consumer protection, nuclear waste protection, financial products, and housing. But the expertise to address the challenges of technology (which represents 7 percent of GDP and a substantially higher growth rate) is spread haphazardly across multiple agencies and departments. One hand rarely knows what the other is doing. Do we need to establish a Digital Protection Agency?

It is time to build a coherent governmental structure to help our democracy address the challenges of data protection, data ownership, privacy, digital citizenship, and platform regulation. Democracy needs trusted guides to help lead it through the twisting and highly technical labyrinth of the digital era.

Apply "societal seat belts" with courage and alacrity: In retrospect, it was a travesty that the United States waited decades to mandate safety glass and seat belts in automobiles. It was also a travesty that auto industry participants fought those mandates for no other reasons than the time and money it took them to retool.

As leading internet companies disclaim responsibility for postings on their platforms, suggest unregulated currency creations, and artfully ignore the effects of their products on children, democracy will need to firmly require "seat belts." We cannot allow our regulatory institutions and businesses to use time and money as excuses to delay the implementation of protections because, they claim, it is "too hard." Designed intelligently, the necessary "seat belts" should not stifle innovation or destroy progress. We cannot delay their use in cars that are accelerating at unprecedented speed.

Match the speed of response to the speed of the threat: Democracy seems to most easily embrace change in times of crisis. An isolationist United States had the eighteenth-largest army in 1938, only to vastly expand its armed forces after Pearl Harbor in time to shape the outcome of the war and the new world order that would persist for decades afterward.

To date, however, our democratic institutions have failed to recognize the crisis in front of us and to adapt the speed of their engagement to the speed of the threat. One need look no further than Mark Zuckerberg's 2018 appearance before Congress to understand that we have not even entered a period of quality policy debate. The threat is outrunning us.

In business, the digital revolution has required recognition of the need for speed and new tools to address new challenges. Democracy must move beyond show hearings and long-term dependence on the court system to truly engage in a policy debate and to drive immediate action. Whatever one's politics, moreover, the election interference of 2016 and beyond should have spurred a "Pearl Harbor" moment of national reaction.

Partner with business on societal change: We have entered a new era of "post-Friedmanism" in business purpose. Businesses are no longer solely tied to the primacy of shareholders and profits and are redefining their relationship with broader stakeholders. Our own firm has been working to redefine impact investing. We are investing billions of dollars in companies that are providing sustainable profits and demonstrably positive social impact.

With this broad-based shift in business purpose, our democratic institutions should partner with business more aggressively to define, design, and deploy "societal seat belts" in a way that will keep positive technological change on track. In areas like skills training, environmental impact, and accessible education, businesses and democratic institutions can be enthusiastic partners instead of regulatory combatants.

Facilitate and support the building of necessary infrastructure: Democratic government helped shape the nature of the automotive, aeronautical, and communication revolutions by helping to build and manage the collective infrastructure of highways, airports, and airwaves. In this technological revolution, the key infrastructure component is likely to be human capital, and the key needs will revolve around workplace skills.

The rapid rate of change engendered by the digital revolution has outstripped the capacity of our traditional education and skills-training infrastructure. Democracy last faced a reskilling task of this magnitude after World War II. The passage of the GI Bill allowed a huge segment of the population, outside of normal school ages, to be retrained. Millions were transformed from soldiers of war to enablers of peacetime prosperity.

Today's low interest rates would allow our government to borrow to invest in rebuilding our human capital infrastructure to meet the opportunities of the digital era—just as it borrowed to build the interstate highway system, enabling the automotive era. We have the means and the need to launch a twenty-first-century equivalent to the GI Bill to retrain and renew our workforce to create economic opportunities so vital to US democracy.

The digital revolution is packed with promise and opportunity. Yet like any major technological advance, its promise will be paired with risk and challenge. New mindsets, new urgency, and new tools will enable both business and democracy to survive and flourish.

Are Facebook, Google, and Amazon the New Tobacco Industry?

JEFF GOODBY

Jeff Goodby is a writer, artist, director, and founder of his advertising agency, Goodby, Silverstein & Partners.

You don't even know me, but you should be angry with me.

I will explain.

As an advertising person, I am complicit in the creation of a shocking industrial system that has turned us all into data-spewing value bots, has undermined elections, and even threatens to curtail our innate creativity as humans. I truly believe that, barring big changes, we may one day look back on the companies responsible for all this with the loathing we feel for the tobacco industry. I know it sounds lame, but I didn't even know this was happening as I artistically and quite successfully pushed the whole thing forward.

Let me back up a bit.

I have always thought of advertising as a corporately funded exploration of who we really are—what we love, what we fear, what we find beautiful, and our deepest desires. It is a kind of group art that we tolerate because it sometimes charms us, educates us, brings us news, and makes beer taste better than it actually does. Beginning many decades ago, data were collected in service of this mission. As consumers, we were of course studied endlessly for our instincts and reactions. Research was gathered face-to-face and in quantitative surveys. We were subtly (and sometimes not so subtly) led toward selecting Tide over Cheer, Mercedes over Audi.

Did these transactions change us? I think they did. They began feeding back versions of ourselves that were suddenly more concerned about beauty, elegance, speed, just doing it, and convenience, and in the process perhaps made us care about issues we wouldn't have otherwise considered. The celebrated advertising creative director Hal Riney groused about this in 1982. "The beauty and whimsy, the cleverness and the suggestion seem to be gone from everything," he wrote. "And they've been replaced by two people holding up a product they would never hold up; and talking about it in a way no one ever talked; and being astonished, pleased, delighted, or surprised about characteristics of a product which in real life would actually rate no more than a grunt, at best."

As computing power increased, data-collecting devices entered our homes and cars and clothing and public spaces. My advertising industry, and especially the retailing arm of it, suddenly had mountains of data about our audiences. We discovered that we could inflame, in great detail, the desire for certain products. More than ever, in virtually invisible ways (*Pokémon Go* was one of the greatest data collectors in history), we treated people as machines who could be manipulated to predictably buy certain brands. Data that once described "types" of consumers were suddenly able to predict individual behavior (or at least the purveyors of such data claimed they could).

Unsuspecting consumers often say that they don't mind trading their data for all the maps, photo sharing, and self-expression that the internet seems to provide for free. In reality, of course, the value of the exchange is shockingly stacked against the user.

The key word in all this of course is *unsuspecting.* In the twentieth century, we could largely tell when we were being advertised to—we could feel the manipulation. It happened in commercial pods, on billboards, in high-gloss magazines. But social media sites changed all that in ways that amounted to a true tectonic shift. Suddenly, we *wanted* to ride the slipstream of this avalanche. We saw it as sharing photos, funny memes, cheap music, and cat videos.

Little did we know, it was so much more than that.

Harvard professor Shoshana Zuboff has termed all of this "surveillance capitalism." It is the business of studying us, without our knowledge, and using such findings to affect our actions. And it doesn't just change our behavior. It changes *us*, in the biggest sense.

In the internet age, the feedback loop created by this studying has a tendency to reduce us to our most monetizable aspects. For example, we become "outdoorsy" because the ads we receive tell us that. We are "naturally beautiful" but not "glamorous" because that's how we are reflected in the ads that reach us. Advertisers tell us it's okay to be "ruthlessly ambitious" or alternatively "rejecting of all ambition" because both of these could be attractive online stances that stimulate sales.

I guess I was okay with the idea of data manipulating the choice of Dawn over Palmolive dish soap, although my online life became flooded with ads for things I had just bought or was thinking about buying. Advertising agencies are really good at anticipating behavior and taking full credit for sales—a person browses the BMW website, so quick, send them an ad! It irked me that digital advertising seemed to rely more on repetition and placement than on wit or charm and was thus an environmental and mental eyesore. Like everyone else, I thought I had trained myself to ignore it all.

However, when my advertising environment seemed to actually be shaping who I was and how I thought of myself, even limiting my creativity, it was crossing some big lines.

It has struck me that the thing we call "creativity" is a matter of making connections between quite disparate things. The mathematician Elwyn Berlekamp once told me, "Inspiration is overrated. With enough information, patterns emerge." T. S. Eliot, too, thought of poetry as the fusing together of diverse things under high pressure. Creativity, in short, is about observation, collection, and drawing conclusions. Our online life, in the long run, has a way of limiting this process. It invisibly controls the flow of information to us. In this way, the data we are fed staunches our creative connection-making.

In the political arena, data manipulation is especially dangerous. Data points, of course, are not inherently evil. They save lives every day in the world's hospitals, make our roads safer, help create food that's fresher and more nutritious, and connect us to centuries of knowledge. But in a political context, things can take a sinister turn. The 2019 documentary film *The Great Hack* traces how the firm Cambridge Analytica undermined the US and British election processes through tactics that may one day be looked back upon as similar to the strategies of Big Tobacco. It traces the tens of thousands of data points about each of us that have been collected, packaged, and sold by the big internet companies. It then outlines how these data were used to send highly targeted, upsetting, even racist messages to unsuspecting "persuadables" and influenced critical 2016 elections. Whether Cambridge Analytica's application of these techniques directly changed the outcomes of the 2016 presidential election and the Brexit vote—people formerly in the company firmly believe it did—it is indisputable that they divided populations in deep ways that will resonate for years. The conflicting narratives are strong and searingly reinforcing.

Once again, we learned–and may be about to learn again in 2020–that stories hold audiences, and held audiences sell more advertising. Complex personality profiles were created from people's whereabouts, their choices of clothes and cars, whether they owned hunting licenses, and analysis of their Twitter habits and Facebook friends. On the basis of these profiles, key audiences were identified and singled out. Innocuous, even charming stories were used to draw those groups in and earn trust. Then more dangerous stories were delivered that, it was determined, would resonate with these audiences–an unbranded ad about Hillary Clinton's alleged abuse of email, say, or something depicting a deluge of criminal immigrants.

And so we come full circle back to me, the advertising guy.

Today's advertising industry is obliviously riding consumer data like a four-year-old on a mechanical bull. In the process, we have no idea how much damage we're doing to our clients and consumers. The worst part is, there is no upside compelling us to care. We are banking billions.

Ironically, advertising is also the force most likely to look Silicon Valley in the eye and extract change. We are the source of the internet's burgeoning income. We have a responsibility to band together to demand more openness and action from the big platforms and to teach people how to more easily turn on privacy settings. As advertisers, our customers simply have a right to know more about what's happening to us all.

It won't be easy. Facebook CEO Mark Zuckerberg's testimony before Congress in 2018 raised serious concerns about the platform's moral compass. Professor Zuboff describes the attitude of Big Data royalty as "radical indifference." There is just too much money involved for them to care.

Government has been slow to get involved. I suspect our representatives don't really even understand the issues. Advertising and marketing budgets could well be the only big stick.

From my standpoint, it may seem altruistic, but it's not. There is actually something in this for people like me. We in advertising used to entertain and charm customers to build brands. In an environment of surveillance capitalism, we are in the process of making the art in advertising less and less necessary and relevant. Without a big course correction, our customers (and voters) will end up hating and distrusting *all* our messages.

So besides the basic human and moral victories from reining in social media, there are clear business benefits. This stuff is threatening the very existence of advertising as we know it, and while you may not find that as dismaying as I do, you should certainly be concerned about the concurrent obfuscation and dumbing down of commerce in general.

What can advertising do? A lot, I think.

For one, we as advertisers can decisively act when advertising is placed on objectionable sites by big media companies who buy advertising placement with profiling algorithms. The tiny stealth organization Sleeping Giants alerted big, general-audience advertisers to the fact that their media buy was resulting in appearances in places like Breitbart News and the *Daily Stormer.* The sponsors immediately requested that this kind of placement stop. Almost overnight, a significant part of Brietbart's income dried up. In fact, Steve Bannon himself has admitted that this is why they went from a paid advertising site to a subscription-based model. This was not an isolated incident. The act had real consequences. Despite death threats and insidious comments, Sleeping Giants continues its good and effective work.

When you think about it, the internet has allowed *anyone* out there to do exactly what Sleeping Giants did. As advertisers and readers, we can all complain to the sponsor when we see racist or homophobic advertising on legitimate websites or legitimate advertising on extreme sites. Unlike Facebook, most advertisers respond quickly and with great sensitivity.

Beyond that, we should immediately outlaw all political advertising in any form, in any media. Naturally, this is a tough one because social media channels, political advertisers, and advertising companies are almost all against such a move. Sure, we can ban paid political announcements from candidates and political action committees. But what about a very convincing, well-produced message of support from a private citizen that gets shared thousands, even millions of time? What is stopping paid political advertising from migrating to private individuals?

But advertising agencies and media buyers could be where the end of political advertising actually gains traction. It's obviously going to be harder to convince social media and political advertising firms to go along with such a ban, but general market advertisers could make it happen simply by swinging their digital advertising toward channels that ban political ads. Modern media planning can do this. Look: Twitter has embraced such a stance. We should reward them for it. Social media companies could additionally stop the microtargeting of anything political. Anything. As soon as someone is paying to digitally disseminate a message, it should be scrutinized for political content.

Sacha Baron Cohen has suggested that social channels be treated as publishers, movie studios, or TV channels that are all either voluntarily or legally restricted in clear ways. Think of the restrictions on magazines like *Hustler*, the rating of movies, and the television prohibitions against cursing and nudity. Cohen wonders why such social media companies aren't internally or even governmentally checked. A seven-second delay before things are posted globally, such as the one practiced by live TV and radio, could help with such regulation. Beyond political advertising, it could even help spot racists and pedophiles.

Finally, short of the Great Facebook Unbooking boycott that we all sometimes dream of undertaking (until a wedding or birth sidetracks us), I have to ask a simple question: Can't agencies and advertisers simply avoid the social sites most resistant to responsible self-regulation? This would, I think, start with Facebook and Google. I truly believe there will be a turning point at which consumers will actively *prefer* brands that don't support those sites. Am I crazy? If Mark Zuckerberg's performances at the Capitol continue to dodge and deflect, I don't think so.

Yes, I'd like to apologize for the behavior of my industry so far in this arena. But as I said, in financial ways, we hold all the cards. In the future, I intend to do everything I can to see that we play them.

ACKNOWLEDGMENTS

This book is a collection of pieces by some of the most intelligent, talented people I know. It was a collaborative effort from day one, and I am deeply grateful to all the contributors for their insights, generosity, and unique perspectives. I feel humbled by the quality of their writing and intellect, as well as the depth and creativity of their minds. Every one of these contributors is on the right side of history.

A number of people deserve special credit for this book. The list starts, first and foremost, with my terrific and talented colleague, Susan Wels, without whom this book would likely never have happened. Susan's fine editing skills are matched only by her thoughtfulness, patience, kindness, and persistence. Her extraordinary editorial skills and wise counsel are present throughout the entire book. Susan has worked with me on all of my books and has been a friend for more than forty years. I simply cannot thank her enough for being such a terrific friend and partner.

Several of my colleagues at Common Sense also played very important roles in making this book happen. My great friend and policy partner in crime, the uniquely talented Bruce Reed, helped shape the book's structure, as well as my own personal contributions. "Cousin Brucie" is one of the best and smartest people I know, and his world-class intellect and policy expertise were invaluable.

My chief of staff, Wilson Tong, played a key role in organizing many aspects of the process, and he helped shape the contours and structure of the book—even while on paternity leave with his infant son, Grayson. Similarly, my talented executive assistant, Saira Malik, helped herd all the cats to make this book come together, and she did so with her customary wisdom and positive outlook on life. Other Common Sense colleagues—including Ellen Pack,

Eliza Panicke, Colby Zintl, Marlene Saritzky, David Kuizenga, Linda Burch, and Lisa Cohen—all also provided much needed input and support.

Chronicle Books was my first and only stop when I conceived of this book. My great friend of thirty years, the fabulous Nion McEvoy—CEO at Chronicle—was my partner from day one, and he greenlit the book in our very first meeting. Our editor, Eva Avery, brought her laid-back and insightful Hawaiian sensibility to every step of the process, and she did an excellent job in helping to shape the entire book. Mark Tauber and Tyrell Mahoney were supportive of the project from the outset, and I owe them and all the fine folks at Chronicle much appreciation for their kindness and support. They have been a great team to work with.

So many of the individual contributors to this book are superstars in their own right and generously donated their talents and wisdom through their writing and personal friendship. A handful of these friends—including Roger McNamee, Sacha Baron Cohen, Marc Benioff, and Willow Bay—have also helped shape the related advocacy and research efforts that led me to write this book, and I am grateful for their ongoing insights and inspiration. I also owe thanks to my friend Jennifer Parker, whose help was instrumental in securing several contributions to the book.

Finally, you simply can't write a book without the love, support, and patience of your family. I dedicated the book to my wife, Liz, who has put up with and supported me for nearly thirty years. But I am also eternally grateful to our fabulous four children—Lily, Kirk, Carly, and Jesse—for being absolutely the best part of my life. I love and cherish each of them beyond words. I am a very lucky man indeed to have such a wonderful family and great group of friends! They all share a piece of this book.

–James P. Steyer

ENDNOTES

INTRODUCTION BY JAMES P. STEYER

Page xiii, “They even attacked us for criticizing the company’s efforts”: https://www.wired.com/story/facebook-for-6-year-olds-welcome-to-messenger-kids/amp.

Page xiii, “A 2019 Pew Research Center survey found that only 50 percent”: Carroll Doherty and Jocelyn Kiley, “Amerians Have Become Much Less Positive about Tech Companies’ Impact on the US,” Pew Research Center, July 29, 2019, https://www.pewresearch.org/fact-tank/2019/07/29/americans-have-become-much-less-positive-about-tech-companies-impact-on-the-u-s.

Page xvii, “According to Ken Auletta, the renowned media”: https://www.mobilemarketer.com/news/googles-q4-ad-revenue-rises-20-as-its-pricing-power-erodes/547666/; https://investor.fb.com/investor-news/press-release-details/2019/Facebook-Reports-Fourth-Quarter-and-Full-Year-2018-Results/default.aspx.

"TECH, HEAL THYSELF" BY ELLEN PAO

Page 32, “After the stock market dropped in 2008, greed turned into layoffs”: Matt Taibbi, “The Great American Bubble Machine,” *Rolling Stone*, April 5, 2010, https://www.rollingstone.com/politics/politics-news/the-great-american-bubble-machine-195229.

Page 34, “In 2013, Tracy Chou started a project to count women engineers”: Tracy Chou, “Where Are the Numbers?” *Medium*, October 11, 2013, https://medium.com/@triketora/where-are-the-numbers-cb997a57252.

Page 35, “Its 2018 report showed that the percentage of Black”: Zoë Bernard, “Silicon Valley’s Most and Least Diverse Venture Capital Firms,” *The Information*, April 18, 2019, https://www.theinformation.com/articles/silicon-valleys-most-and-least-diverse-venture-capital-firms.

Page 35, “According to Richard Kerby’s research in 2018, 82 percent”: Richard Kerby, “Where Did You Go to School?” *Medium*, “Noteworthy—the Journal Blog,” July 30, 2018.

Page 35, “Also in 2020, All Raise reported that 65 percent”: These details come from the following: Pam Kostka, “More Women Became VC Partners Than Ever Before in 2019, but 65% of Venture Firms Still Have No Women Partners,” *Medium*, “All Raise,” February 7, 2020; and Pitchbook and All Raise, “All In: Women in the VC Ecosystem,” November 12, 2019.

Page 35, “In 2017, according to a RateMyInvestor report, only 1 percent”: RateMyInvestor, “Diversity in US Startups,” ratemyinvestor.com/diversity_report.

Page 35, "As a result, the percentages of wealth generated by women": Emily Kramer, "Fair Equity Is Table Stakes: 2019 Gender Equity Gap Study," *Carta* (blog), November 4, 2019.

Page 36, "The numbers on harassment in the tech industry are damning": The two surveys in this paragraph are "State of Startups 2017," First Round, http://stateofstartups.firstround.com/2017; and "Survey of YC Female Founders on Sexual Coercion and Assault by Angel and VC Investors," *Y Combinator* (blog), October 15, 2018, https://blog.ycombinator.com/survey-of-yc-female-founders-on-sexual-harassment-and-coercion-by-angel-and-vc-investors.

Page 37, "Today, CEOs and founders have changed their minds about their social": The quotes in this paragraph come from the following: Evan Williams in David Streitfeld, "'The Internet Is Broken': @ev Is Trying to Salvage It," *New York Times*, May 20, 2017; 8chan founder in Timothy McLaughlin, "The Weird, Dark History of 8chan," *Wired*, August 6, 2019; and Dick Costolo in Nitasha Tiku and Casey Newton, "Twitter CEO: 'We suck at dealing with abuse,'" *The Verge*, February 4, 2015.

Page 38, "In November 2019, in the United Kingdom, several female members": Megan Specia, "Threats and Abuse Prompt Female Lawmakers to Leave UK Parliament," *New York Times*, November 1, 2019.

Page 38, "In March 2020, Amazon and Instacart employees went on": Nitasha Tiku and Jay Greene, "Workers Protest at Instacart, Amazon, and Whole Foods for Health Protections and Hazard Pay," *Washington Post*, March 30, 2020.

Page 38, "Salesforce, Microsoft, Google, Palantir, and Amazon employees have objected": https://www.businessinsider.com/github-employees-ice-contracts-protest-microsoft-2019-11 and https://www.cnbc.com/2019/08/19/google-employees-implore-leaders-to-stop-working-with-us-bcp-ice.html.

Page 38, "Google and Facebook faced broad criticism when they considered": Will Knight, "The Value and Ethics of Using Phone Data to Monitor COVID-19," *Wired*, March 18, 2020.

Page 39, "Jeff Bezos has the highest net worth, at over $100 billion": Giacomo Tognini, "World's 20 Richest, Led by Jeff Bezos, Shed More Than $78 Billion Amid Thursday's Market Rout," *Forbes*, March 12, 2020.

"WE NEED A NEW CAPITALISM, BASED ON TRUST" BY MARC BENIOFF

Page 45, "Research from JUST Capital found that companies that do best": Hernando Cortina, "JUST Business, Better Margins," JUST Capital, June 2019, https://justcapital.com/reports/just-business-better-margins.

Page 45, "LRN, a leading ethics and compliance company, surveyed": LRN Corporation, "Vast Majority of Employees Say There's an Urgent Need for Moral Leadership in Business Today—but It's All Too Rare, New Report Says," February 19, 2019, https://www.globenewswire.com/news-release/2019/02/19/1734464/0/en/Vast-Majority-of-Employees-Say-There-s-an-Urgent-Need-for-Moral-Leadership-in-Business-Today-But-It-s-All-Too-Rare-New-Report-Says.html.

Page 47, "A 2018 article in the *Harvard Business Review* perhaps summed it": Aaron K. Chatterji and Michael W. Toffel, "Divided We Lead," *Harvard Business Review*, March 2018, https://hbr.org/cover-story/2018/03/divided-we-lead.

Page 50, "According to the World Economic Forum's 2018 *The Future of Jobs Report*": World Economic Forum, *The Future of Jobs Report 2018*, http://www3.weforum.org/docs/WEF_Future_of_Jobs_2018.pdf.

Page 50: "In the US there are more than 600,000 open tech jobs:" The number of current open computing jobs comes from the sum of the per-state jobs data from the Conference Board Help Wanted OnLine service, https://www.conference-board.org/data/helpwantedonline.cfm. For more information, see the Conference Board (www.conference-board.org/us) and the National Center for Education Statistics (www.nces.ed.gov).

"MOST TECH COMPANIES DON'T CARE" BY NICHOLAS D. KRISTOF AND SHERYL WuDUNN

Page 57, "In 2018, some 130 of the biggest law firms devoted more than": "A First: Law Firm Pro Bono Annual Hours Exceed Five Million," Pro Bono Institute, June 13, 2019, http://thepbeye.probonoinst.org/2019/06/13/a-first-law-firm-pro-bono-hours-exceed-five-million.

Page 58, "In 2019, its new CEO, Alan Jope, said Unilever might shed products": P. J. Bednarski, "Unilever Chief Says Some of Its Brands without 'Purpose' May Have to Go," *MediaPost*, July 26, 2019, https://www.mediapost.com/publications/article/338664/unilever-chief-says-some-of-its-brands-without-pu.html.

"TECHNOLOGY AND SOCIAL CONNECTION" BY VIVEK MURTHY

Page 65, "According to a 2018 report by the Henry J. Kaiser Family Foundation": Bianca DiJulio, Liz Hamel, Cailey Muñana, and Mollyann Brodie, "Loneliness and Social Isolation in the United States, the United Kingdom, and Japan: An International Survey," Henry J. Kaiser Family Foundation, August 30, 2018, https://www.kff.org/other/report/loneliness-and-social-isolation-in-the-united-states-the-united-kingdom-and-japan-an-international-survey.

Page 65, "A 2018 AARP study using the rigorously validated UCLA loneliness": G. Oscar Anderson and Colette E. Thayer, "Loneliness and Social Connections: A National Survey of Adults 45 and Older," AARP Research, "Life and Liesure," September 2018, http://doi.org/10.26419/res.00246.001.

Page 65, "And in a 2018 national survey by the US health insurer Cigna": "2018 Cigna US Loneliness Index: Survey of 20,000 Americans Examining Behaviors Driving Loneliness in the United States," Cigna, May 2018, https://www.multivu.com/players/English/8294451-cigna-us-loneliness-survey/docs/IndexReport_1524069371598-173525450.pdf.

Page 65, "Dr. Julianne Holt-Lunstad of Brigham Young University": Julianne Holt-Lunstad, Timothy Smith, and J. Bradley Layton, "Social Relationships and Mortality Risk: A Meta-Analytic Review," *PLOS Medicine* 7, no. 7 (July 2010), https://doi.org/10.1371/journal.pmed.1000316.

Page 66, "Studies also suggest that lonely people are more likely to have": Louise C. Hawkley and John T. Cacioppo, "Loneliness Matters: A Theoretical and Empirical Review of Consequences and Mechanisms," *Annals of Behavioral Medicine* 40, no. 2 (October 2010): 218–27, https://doi.org/10.1007/s12160-010-9210-8.

Page 67, "UK researchers Andrew Przybylski and Netta Weinstein found that": Andrew K. Przybylski and Netta Weinstein, "A Large-Scale Test of the Goldilocks Hypothesis: Quantifying the Relations Between Digital-Screen Use and the Mental Well-Being of Adolescents," *Psychological Science* 28, no. 2 (January 2017), https://doi.org/10.1177%2F0956797616678438.

Page 67, "Recent estimates, however, put the average time teenagers": Vicky Rideout, "The Common Sense Census: Media Use by Tweens and Teens," Common Sense, 2015, https://www.commonsensemedia.org/sites/default/files/uploads/research/census_researchreport.pdf.

Page 69, "As MIT neuroscientist Dr. Earl Miller explained in a 2008 interview": Jon Hamilton, "Think You're Multitasking? Think Again," NPR, October 2, 2008, https://www.npr.org/templates/story/story.php?storyId=95256794.

Page 70, "As Przybylski and Weinstein found in their experiments": Andrew K. Przybylski and Netta Weinstein, "Can You Connect with Me Now? How the Presence of Mobile Communication Technology Influences Face-to-Face Conversation Quality," *Journal of Social and Personal Relationships* 30, no. 3 (July 2012): 237–46, https://doi.org/10.1177/0265407512453827.

"KIDS INTERRUPTED: HOW SOCIAL MEDIA DERAILS ADOLESCENT DEVELOPMENT" BY MADELINE LEVINE

Page 79, "As compared with adolescent boys, adolescent girls use": Jean M. Twenge, *iGen: Why Today's Super-Connected Kids Are Growing Up Less*

Rebellious, More Tolerant, Less Happy—and Completely Unprepared for Adulthood (New York: Atria Books, 2017).

Page 84, "We also know that over half of teens describe themselves as being": Twenge, *iGen*.

"WHY SECTION 230 HURTS KIDS AND WHAT TO DO ABOUT IT" BY BRUCE REED AND JAMES P. STEYER

Page 95, "According to former California Representative Chris Cox, who wrote Section 230": Alina Selyukh, "Section 230: A Key Legal Shield for Facebook, Google Is About to Change," NPR, March 21, 2018, https://www.npr.org/sections/alltechconsidered/2018/03/21/591622450/section-230-a-key-legal-shield-for-facebook-google-is-about-to-change.

Page 97, "A Pew Research Center survey found that four out of five": Aaron Smith, Skye Toor, and Patrick Van Kessel, "Many Turn to YouTube for Children's Content, News, How-To Lessons," Pew Research Center, Novermber 7, 2018, https://www.pewresearch.org/internet/2018/11/07/many-turn-to-youtube-for-childrens-content-news-how-to-lessons.

Page 97, "As the channel's proud parent": *Federal Trade Commission v. Google LLC and YouTube LLC*, "Complaint for Permament Injunction," 2019, https://www.ftc.gov/system/files/documents/cases/youtube_complaint.pdf.

Page 97, "YouTube videos aimed at kids have shown all manner of violence": For more on these examples, see the following: Alun Palmer and Ben Griffiths, "Assault 'N Peppa: Kids Left Traumatized after Sick YouTube Clips Showing Peppa Pig Characters with Knives and Guns Appear on App for Children," *Sun*, July 9, 2016, https://www.thesun.co.uk/news/1418668/kids-left-traumatised-after-sick-youtube-clips-showing-peppa-pig-characters-with-knifes-and-guns-appear-on-app-for-children; Russell Brandom, "Inside Elsagate, the Conspiracy-Fueled War on Creepy YouTube Kids Videos," *Verge*, December 8, 2017, https://www.theverge.com/2017/12/8/16751206/elsagate-youtube-kids-creepy-conspiracy-theory; and Nick Statt, "YouTube Bans Account of Parents Whose Prank Videos Depicted Child Abuse," *Verge*, July 18, 2018, https://www.theverge.com/2018/7/18/17588668/youtube-familyoffive-child-neglect-abuse-account-banned.

Page 97, "An exhaustive research study funded by the European Union": Kostantinos Papadamou et al., "Disturbed YouTube for Kids: Characterizing and Detecting Inappropriate Videos Targeting Young Children" (January 2019), https://arxiv.org/pdf/1901.07046.pdf.

Page 98, "As technology writer James Bridle warned in 2017": James Bridle, "Something Is Wrong on the Internet," *Medium*, November 6, 2017, https://medium.com/@jamesbridle/something-is-wrong-on-the-internet-c39c471271d2.

Page 99, "Hunters and gun owners don't benefit from that law": https://news.guns.com/wp-content/uploads/2018/09/NSSF-MSR-Production-Estimates-2017.pdf.

Page 99, "Instead of putting up real guardrails against hate speech": Casey Newton, "The Trauma Floor," *Verge*, February 25, 2019, https://www.theverge.com/2019/2/25/18229714/cognizant-facebook-content-moderator-interviews-trauma-working-conditions-arizona.

Page 99, "As Jeff Kosseff, the law's self-proclaimed biographer": Adi Robertson, "Why the Internet's Most Important Law Exists and How People Are Still Getting It Wrong," *Verge*, June 21, 2019, https://www.theverge.com/2019/6/21/18700605/section-230-internet-law-twenty-six-words-that-created-the-internet-jeff-kosseff-interview.

Page 100, "In their article 'The Internet Will Not Break' ": Danielle Citron and Benjamin Wittes, "The Internet Will Not Break: Denying Bad Samaritans Secion 230 Immunity," *Fordham Law Review*, July 24, 2017, https://papers.ssrn.com/sol3/papers.cfm?abstract_id=3007720.

Page 100, "Citron argues that courts should ask whether providers have": Danielle Citron, "Tech Companies Get a Free Pass on Moderating Content," *Slate*, October 16, 2019, https://slate.com/technology/2019/10/section-230-cda-moderation-update.html.

Page 101, "She challenged leaders of nations and corporations around the world": Jacinda Ardern, "Jacinda Ardern: How to Stop the Next Christchurch Massacre," *New York Times*, May 11, 2019, https://www.nytimes.com/2019/05/11/opinion/sunday/jacinda-ardern-social-media.html.

"USING TECHNOLOGY TO BOOST KIDS' BRAIN DEVELOPMENT" BY CHELSEA CLINTON

Page 106, "Further, children as early as two are spending an average": Victoria Rideout, "The Common Sense Census: Media Use by Kids Age Zero to Eight," Common Sense Media, 2017, https://www.commonsensemedia.org/sites/default/files/uploads/research/csm_zerotoeight_fullreport_release_2.pdf.

"BOLSTERING DEMOCRACY'S IMMUNE SYSTEM" BY CRAIG NEWMARK

Page 123, "Since 2015, bad actors have been weaponizing information technology": The details in this paragraph come from the following: Rob Barry, "Russian Trolls Tweeted Disinformation Long Before US Election," *Wall Street Journal*, February 20, 2018, https://www.wsj.com/graphics/russian-trolls-tweeted-disinformation-long-before-u-s-election; Matthew Hindman and Vladimir Barash, "Disinformation, 'Fake News,' and Influence Campaigns on

Twitter," Knight Foundation (October 2018), https://knightfoundation.org/features/misinfo; Joe Sonka, "Thousands of Twitter 'Bots' Targeted Kentucky with Fake News on Election Night," *USA Today*, November 11, 2019, https://www.usatoday.com/story/news/politics/2019/11/11/kentucky-elections-2019-thousands-twitter-bots-spread-fake-facts/2564439001; and Emily Stewart, "Facebooks Has Taken Down Billions of Fake Accounts, but the Problem Is Still Getting Worse," *Vox*, May 23, 2019, https://www.vox.com/recode/2019/5/23/18637596/facebook-fake-accounts-transparency-mark-zuckerberg-report.

Page 124, "Outdated voting technology and technology that does not leave": Lawrence Norden and Andrea Córdova McCadney, "Voting Machines at Risk: Where We Stand Today," Brennan Center for Justice, March 5, 2019, https://www.brennancenter.org/our-work/research-reports/voting-machines-risk-where-we-stand-today.

Page 124, "While adversaries wage an information war on us and threaten to tamper": The details in this paragraph come from the following: Brennan Center for Justice, "New Voting Restrictions in America," October 1, 2019 (updated November 19, 2019), https://www.brennancenter.org/our-work/research-reports/new-voting-restrictions-america; and Ankita Rao et al., "Is America a Democracy? If so, Why Does It Deny Millions the Vote?," *Guardian*, November 7, 2019, https://www.theguardian.com/us-news/2019/nov/07/is-america-a-democracy-if-so-why-does-it-deny-millions-the-vote.

Page 124, "They are working to foster relationships with social media": Andy Greenberg, "Facebook's Ex-Security Chief Details His 'Observatory' for Internet Abuse," *Wired*, July 25, 2019, https://www.wired.com/story/alex-stamos-internet-observatory.

Page 124, "Adversaries work to get mainstream media—the source": Victoria Kwan, "Responsible Reporting in an Age of Information Disorder," First Draft (October 2019), https://firstdraftnews.org/wp-content/uploads/2019/10/Responsible_Reporting_Digital_AW-1.pdf?x88639.

Page 125, "Ahead of the 2020 US presidential election, the organization": David Bauder, "Craigslist Founder Donates to Group Fighting Fake News," *Associated Press*, October 24, 2019, https://apnews.com/a54fc1be943942f8aad614529d46a7d1.

Page 125, "Dan Gillmor—cofounder and director of the News Co/Lab": Dan Gillmor, "Dear Journalists, Stop Being Loudspeakers for Liars," *Medium*, June 15, 2018, https://medium.com/@dangillmor/dear-journalists-stop-letting-liars-use-your-platforms-as-loudspeakers-cc64c4024eeb.

Page 125, "Tactics may include adopting the 'truth sandwich' ": Margaret Sullivan, "Instead of Trump's Propaganda, How about a Nice 'Truth Sandwich'?" *Washington Post*, June 17, 2018, https://www.washingtonpost.com/lifestyle/style/instead-of-trumps-propaganda-how-about-a-nice-truth-sandwich/2018/06/15/80df8c36-70af-11e8-bf86-a2351b5ece99_story.html.

Page 125, "To that end, ongoing work at the Columbia Journalism School" Information on these programs comes from the following: "Newmark Philanthropies Awards $15 Million to Bolster Journalism Ethics," *Philanthropy News Digest*, February 11, 2019, https://philanthropynewsdigest.org/news/newmark-philanthropies-awards-15-million-to-bolster-journalism-ethics; and Richard Danielson, "Craigslist's Craig Newmark Gives Poynter $5 Million for Ethics Center," *Tampa Bay Times*, February 6, 2019, https://www.tampabay.com/business/craiglists-craig-newmark-gives-poynter-5-million-for-ethics-center-20190206.

Page 126, "Equally important, we need to help restore a vanishing local press": Details in this paragraph come from the following: Penelope Muse Abernathy, "The Loss of Newspapers and Readers," in "The Expanding News Desert," Center for Innovation and Sustainability in Local Media, Hussman School of Journalism and Media (UNC, 2018), https://www.usnewsdeserts.com/reports/expanding-news-desert/loss-of-local-news/loss-newspapers-readers; Mike Scutari, "The American Journalism Project Is Poised to Allocate Millions," Inside Philanthropy, April 1, 2019, https://www.insidephilanthropy.com/home/2019/4/1/the-american-journalism-project-is-poised-to-allocate-millions-where-will-the-money-go; and Sara Fischer, "The Race to Fill Local Newsrooms," *Axios*, December 2, 2019, https://www.axios.com/race-to-fill-local-newsrooms-28d95883-1378-4301-9263-ff4f3590c395.html.

"RECLAIMING DEMOCRACY" BY MARIETJE SCHAAKE

Page 130, "Alibaba's cofounder Jack Ma makes an explicit case for corporations": Organización Mundial del Comercio, "*En Una Sesión de Alto Nivel se Destaca el Potencial del Comercio Electrónico Como Motor del Crecimiento y la Inclusión*," October 4, 2018, https://www.wto.org/spanish/news_s/news18_s/pf18_04oct18_s.htm.

"THE ASSAULT ON CIVIL DISCOURSE AND AN INFORMED ELECTORATE" BY SENATOR MARK WARNER

Page 136, "In 1791, James Madison asserted that 'whatever facilitates a general' ": James Madison, "Public Opinion," *National Gazette* (December 1791).

Page 137, "While I am encouraged that governments around the world": See "General Data Protection Regulation (EU) 2016/679," *Official Journal of the*

European Union (2016); and Mark Scott, "Want to Regulate Big Tech? Sorry, Europe Beat You to It," *Politico Europe*, April 11, 2019.

Page 138, "The bottom line is, the United States was unprepared": "Report of the Select Committee on Intelligence, United States Senate on Russian Active Measures Campaigns and Interference in the 2016 U.S. Election," Report 116-XX, 116th Congress, Select Committee on Intelligence, United States Senate (2019).

Page 138, "Gamergate was a concerted harassment campaign waged": Caitlin Dewey, "The Only Guide to Gamergate You Will Ever Need to Read," *Washington Post*, October 14, 2014.

Page 138, "Already we have seen this Russian playbook expand": The details here come from the following: Paul Mozur, "A Genocide Incited on Facebook, with Posts from Myanmar's Military," *New York Times*, October 15, 2018; Brenda Goh, "'All the Forces': China's Global Social Media Push Over Hong Kong Protests," *Reuters*, August 22, 2019; Lauren Etter, "What Happens When the Government Uses Facebook as a Weapon?" *Bloomberg Businessweek*, December 7, 2017; and Declan Walsh and Nada Rashwan, "'We're at War': A Covert Social Media Campaign Boosts Military Rulers," *New York Times*, September 6, 2019.

Page 139, "The truth is: Western companies who help authoritarian": Olivia Solon, "'It's Digital Colonialism': How Facebook's Free Internet Service Has Failed Its Users," *Guardian*, July 27, 2017.

Page 140, "In 2018, Americans were defrauded on Facebook to the tune": Jack Nicas, "Facebook Connected Her to a Tattooed Soldier in Iraq. Or So She Thought," *New York Times*, July 28, 2019.

Page 140, "Neither the defrauded Americans nor the impersonated service": Hemu Nigam, "Facebook's Flaws Let Fake-Love Scammers Bilk Thousands of Victims," *Medium*, August 28, 2019.

Page 141, "Rather than promoting pragmatic rules of the road for the digital economy": See Tony Romm and Drew Harwell, "White House Declines to Back Christchurch Call to Stamp Out Online Extremism Amid Free Speech Concerns," *Washington Post*, May 15, 2019; and David McCabe, "Lawmakers Blast Administration for Tech Shield in Trade Deals," *New York Times*, October 16, 2019.

Page 141, "In 2019, Facebook was caught flat-footed in the face of": Sarah Mervosh, "Distorted Videos of Nancy Pelosi Spread on Facebook and Twitter, Helped by Trump," *New York Times*, May 24, 2019.

Page 142, "Instead, we've seen a number of cases where platforms have worked": See Jeremy Merrill and Ariana Tobin, "Facebook Moves to Block Ad Transparency Tools—Including Ours," *ProPublica*, January 28, 2019; and Peter

Eavis, "Facebook Still Faces Questions About Russia's Reach," *New York Times*, April 9, 2018.

Page 142, "The postal system and the telegraph and the radio were all essential": Richard John, "The Politics of Innovation," *Daedalus* 127, no. 4 (Fall 1998): 187–214.

"THE INFORMALITY MACHINE" BY YUVAL LEVIN

Page 163, "As a result, we sometimes find it hard to really feel like": Daniel Boorstin, *The Image: A Guide to Pseudo-Events in America* (New York: Random House, 1962), 74.

Page 164, "Over time, as we recognize these problems, we may find ways": This point is well argued in Glenn Harlan Reynolds, "Social Media Threat: People Learned to Survive Disease, We Can Handle Twitter," *USA Today*, November 20, 2017.

"THE THIEF IN OUR POCKETS: THE DARK SIDE OF SMART TECH" BY LAURIE SANTOS

Page 168, "In fact, one recent study found that 62 percent of people": Daniel J. Kruger et al., "Cell Phone Use Latency in a Midwestern USA University Population," *Journal of Technology in Behavioral Science* 2 (2017): 56–59, https://link.springer.com/article/10.1007/s41347-017-0012-8.

Page 169, "For instance: 80 percent of Americans say they used a cell phone": For these details, see the following: Lee Rainie and Kathryn Zickuhr, "Americans' Views on Mobile Etiquette," Pew Research Center, August 26, 2015, https://www.pewresearch.org/internet/2015/08/26/americans-views-on-mobile-etiquette; AAA Foundation for Traffic Safety, "2014 Traffic Safety Culture Index," 2015, https://aaafoundation.org/2014-traffic-safety-culture-index; Deborah R. Tindell and Robert W. Bohlander, "The Use and Abuse of Cell Phones and Text Messaging in the Classroom. A Survey of College Students," *College Teaching* 60, no. 1 (2012); and T. Smith, E. Darling, and B. Searles, "2010 Survey on Cell Phone Use while Performing Cardiopulmonary Bypass," *Perfusion* 26, no. 5 (May 2011): 375–80.

Page 169, "Merely hearing a cell phone notification during a pleasant activity": Elif Isikman et al., "The Effects of Curiosity-Evoking Events on Activity Enjoyment," *Journal of Experimental Psychology: Applied* 22, no. 3 (2016): 319–30, https://www.marshall.usc.edu/sites/default/files/ulkumen/intellcont/TheEffectsofCuriosityEvokingEvents-2.pdf.

Page 170, "Business school professor Adrian Ward and his colleagues have found": Adrian F. Ward et al., "Brain Drain: The Mere Presence of One's Own Smartphone Reduces Available Cognitive Capacity," *Journal of the*

Association of Consumer Research 2, no. 2 (April 2017), https://www.journals.uchicago.edu/doi/pdfplus/10.1086/691462.

Page 170, "Psychologist Elizabeth Dunn and her colleagues have a number of new studies": See Ryan Dwyer, Kostadin Kushlev, and Elizabeth Dunn, "Smartphone Use Undermines Enjoyment of Face-to-Face Social Interactions," *Journal of Experimental Psychology* 78 (November 2017), https://www.researchgate.net/publication/320895804_Smartphone_use_undermines_enjoyment_of_face-to-face_social_interactions; and Kostadin Kushlev and Elizabeth Dunn, "Smartphones Distract Parents from Cultivating Feelings of Connection When Spending Time with Their Children," *Journal of Social and Personal Relationships* 36, no. 6 (June 2019): 1,619–39, https://journals.sagepub.com/doi/full/10.1177/0265407518769387.

Page 170, "In one study, Dunn allowed strangers": Kostadin Kushlev et al., "Smartphones Reduce Smiles Between Strangers," *Computers in Human Behavior* 91 (February 2019): 12–16, https://doi.org/10.1016/j.chb.2018.09.023.

Page 171, "The psychologist and author Jean Twenge has argued for a direct causal": Jean M. Twenge, "Have Smartphones Destroyed a Generation?" *Atlantic* (September 2017), https://www.theatlantic.com/magazine/archive/2017/09/has-the-smartphone-destroyed-a-generation/534198/?utm_source=twb.

"TECHNOLOGY CAN AUGMENT OUR HUMANITY OR CONSUME IT" BY ARIANNA HUFFINGTON

Page 179, "The possibility of creating an alternative narrative is one people": Quote by Anne Applebaum from Jessi Hempel, "Social Media Made the Arab Spring, but Couldn't Save It," *Wired*, January 26, 2016, https://www.wired.com/2016/01/social-media-made-the-arab-spring-but-couldnt-save-it.

Page 180, "A 2018 Gallup poll found that 67 percent": Ben Wigert and Sangeeta Agrawal, "Employee Burnout, Part 1: The 5 Main Causes," Gallup, July 12, 2018, https://www.gallup.com/workplace/237059/employee-burnout-part-main-causes.aspx.

Page 180, "A 2019 study by the Cincinnati Children's Hospital Medical Center": John S. Hutton et al., "Associations Between Screen-Based Media Use and Brain White Matter Integrity in Preschool-Aged Children," *JAMA Pediatrics* 174, no. 1 (2020), https://jamanetwork.com/journals/jamapediatrics/article-abstract/2754101.

Page 180, "A 2019 survey by Common Sense Media": Michael B. Robb, "The New Normal: Parents, Teens, Screens, and Sleep in the United States," Common Sense Media, 2019, https://www.commonsensemedia.org/sites

/default/files/uploads/research/2019-new-normal-parents-teens-screens-and-sleep-united-states.pdf.

Page 180, "A 2019 survey by Cigna found that more than half": Cigna, "Loneliness and the Workplace," 2020, https://www.cigna.com/static/www-cigna-com/docs/about-us/newsroom/studies-and-reports/combatting-loneliness/cigna-2020-loneliness-factsheet.pdf.

Page 181, "A study by researchers from Brigham Young University": Selby Frame, "Julianne Holt-Lunstad Probes Loneliness, Social Connections," American Psychological Association, October 18, 2017, https://www.apa.org/members/content/holt-lunstad-loneliness-social-connections.

Page 182, "According to a 2019 report by Global Market Insights": Sumant Ugalmugle and Rupali Swain, "Digital Health Market Size by Technology," Global Market Insights, February 2019, https://www.gminsights.com/industry-analysis/digital-health-market.

"THE NEW JIM CODE" BY RUHA BENJAMIN

Page 211, "Taken together, these images reflect and reinforce popular stereotypes": Angela Bronner Helm, "3 black teens' Google search sparks outrage," *Root*, June 12, 2016, https//www.theroot.com/3-black-teens-google-search-sparks-outrage-1790855635.

Page 211, "The original viral video that sparked the controversy": Jordan Pearson, "It's our fault that AI thinks white names are more 'pleasant' than black names." Motherboard, August 26, 2016, https://motherboard.vice.com/en_us/article/z43qka/its-out-fault-that-ai-thinks-white-names-are-more-pleasant-than-black-names.

Page 212, "According to the company, Google itself uses 'over 200 unique signals'": See "How search algorithms work," https://www.google.co.uk/insidesearch/howsearchworks/algorithms.html.

Page 212, "Or, as one observer put it": See Ethan Chiel, "'Black teenagers' vs. 'White teenagers': Why Google's algorithm displays racist results," *Splinter* News, June 10, 2016. In its own defense, the company explained this: "'Our image search results are a reflection of content from across the web, including the frequency with which types of images appear and the way they're described online,' a spokesman told the *Mirror*. This means that sometimes unpleasant portrayals of sensitive subject matter online can affect what image search results appear for a given query. These results don't reflect Google's own opinions or beliefs—as a company, we strongly value a diversity of perspectives, ideas and cultures." http://splinternews.com/black-teenagers-vs-white-teenagers-why-googles-algor-1793857436.

Page 212, "As Noble reports, the pornography industry has billions of dollars": Siobahn Roberts, "The Yoda of Silicon Valley," *New York Times*, December 17, 2018, https://www.nytimes.com/2018/12/17/science/donald-knuth-computers-algorithms-programming.html.

Page 212, "And so the struggle to democratize information": Sociologist Zeynep Tufekci (2019) puts it thus: "These companies–which love to hold themselves up as monuments of free expression–have attained a scale unlike anything the world has ever seen; they've come to dominate media distribution, and they increasingly stand in for the public sphere itself. But at their core, their business is mundane: They're ad brokers. To virtually anyone who wants to pay them, they sell the capacity to precisely target our eyeballs. They use massive surveillance of our behavior, online and off, to generate increasingly accurate, automated predictions of what advertisements we are most susceptible to and what content will keep us clicking, tapping, and scrolling down a bottomless feed."

Page 213, "Men are shown ads for high-income jobs,": Lauren Kirchner, "When big data becomes bad data," ProPublica, September 2, 2015, https://www.propublica.org/article/when-big-data-becomes-bad-data.

Page 213, "In 2015 Google Photos came under fire": Max Plenke. "Google just misidentified 2 black people in the most racist way possible," Mic Network, June 30, 2015, https://mic.com/articles/121555/google-photos-misidentifies-african-americans-as-gorillas.

Page 213, "Further concerns may arise as AI is given agency": Aylin Caliskan, Joanna J. Bryson, and Arvind Narayanan, "Semantics derived automatically from language corpora contain human-like biases," *Science* 356.6334, 2017, p. 186.

Page 213, "But even after programmers edited the algorithm": Jeffrey Dastin, "Amazon scraps secret AI recruiting tool that showed bias against women," *Reuters*, October 9, 2018, https://www.reuters.com/article/us-amazon-com-jobs-automation-insight/amazon-scraps-secret-a-recruiting-tool-that-showed-bias-against-women-idUSKCN1MK08G.

Page 214, "In fact one HR employee for a major company recommends": Stephen Buranyi, "How to persuade a robot that you should get the job," *Guardian*, March 3, 2018, https://www.theguardian.com/technology/2018/mar/04/robots-screen-candidates-for-jobs-artificial-intelligence.

"INSIDE CULT 2.0" BY RENÉE DiRESTA

Page 234, "We have reached the point when even the Federal Bureau": Asher Stockler, "For First Time, FBI Finds Conspiracy-Backed Crimes Drive

Domestic Terror: Report," *Newsweek*, August 1, 2019, https://www.newsweek.com/conspiracy-trump-qanon-fbi-intelligence-1452236.

Page 234, "The Tree of Life synagogue shooter was active in communities": Anti-Defamation League, "How Conspiracy Theories Can Kill," November 14, 2018, https://www.adl.org/blog/how-conspiracy-theories-can-kill.

Page 235, "What a movement such as QAnon has going for it": Quotes by Rachel Bernstein from interviews with the author.

Page 236, "As Alexis Madrigal writes in the *Atlantic*": Alexis C. Madrigal, "The Reason Conspiracy Videos Work so Well on YouTube," *Atlantic*, February 21, 2019.

"HOW TECHNOLOGY CAN HUMANIZE EDUCATION" BY SAL KHAN

Page 251, "An analysis of nearly three hundred studies on mastery learning": Stephen A. Anderson, "Synthesis of Research on Mastery Learning," *Information Analyses* (November 1, 1994).

Page 251, "When students in the Long Beach Unified School District": Kelli Millwood, Kodi Weatherholtz, and Rajendra Chattergoon, *Use of Khan Academy and Mathematics Achievement: A Correlational Study with Long Beach Unified School District* (Mountain View, CA: Khan Academy, 2019).

"MAKING INTERNET PLATFORMS ACCOUNTABLE" BY ROGER McNAMEE

Page 269, "Myanmar struggled to get Facebook to take action": Paul Mozur, "A Genocide on Facebook, with Posts from Myanmar's Military," *New York Times*, October 15, 2018, https://www.nytimes.com/2018/10/15/technology/myanmar-facebook-genocide.html; Source Vindu Goel, Hari Kumar, and Sheera Frenkel, "In Sri Lanka, Facebook Contends with Shutdown Amid Mob Violence," *New York Times*, March 8, 2018, https://www.nytimes.com/2018/03/08/technology/sri-lanka-facebook-shutdown.html.

Page 271, "They have also experimented with behavioral manipulation": Robert Epstein, PhD, Senior Research Psychologist, American Institute for Behavioral Research and Technology testimony in front of US Senate, "Why Google Poses a Serious Threat to Democracy, and How to End That Threat."

Page 274, "Facebook employees worked side by side with employees": Noah Kulwin, "Trump Campaign Got Help from At Least 8 Silicon Valley Staffers," *Vice*, October 31, 2017, https://www.vice.com/en_us/article/8xmvkg/trump-campaign-had-help-from-8-silicon-valley-staffers.

"THE CHANGE IN THE NATURE OF CHANGE" BY JAMES G. COULTER

Page 294, "In Gould and Eldredge's words, The history of evolution": Niles Eldredge and Stephen Jay Gould, "Punctuated Equilibria: An Alternative to Phyletic Gradualism," in *Essential Readings in Evolutionary Biology*, eds. Francisco J. Ayala and John C. Avise, 82–115 (Baltimore: Johns Hopkins University Press, 2014).